Decoding Large Language Models

An exhaustive guide to understanding, implementing, and optimizing LLMs for NLP applications

Irena Cronin

Decoding Large Language Models

Copyright © 2024 Packt Publishing

Group Product Manager: Niranjan Naikwadi

Publishing Product Manager: Nitin Nainani

Book Project Manager: Hemangi Lotlikar

Senior Editor: Rohit Singh

Technical Editor: Rahul Limbachiya

Copy Editor: Safis Editing

Proofreader: Rohit Singh

Indexer: Manju Arasan

Production Designer: Nilesh Mohite

DevRel Marketing Executive: Vinishka Kalra

First published: October 2024

Production reference: 1101024

Published by Packt Publishing Ltd.

Grosvenor House

11 St Paul's Square

Birmingham

B3 1RB, UK.

ISBN 978-1-83508-465-6

www.packtpub.com

To the memory of my husband, Danny, and his love of all things tech.

– Irena Cronin

Contributors

About the author

Irena Cronin is the SVP of product for DADOS Technology, which is making an Apple Vision Pro data analytics and visualization app. She is also the CEO of Infinite Retina, which provides research to help companies develop and implement AI, AR, and other new technologies for their businesses. Before this, she worked for several years as an equity research analyst and gained extensive experience in evaluating both public and private companies.

Irena has a joint MBA/MA from the University of Southern California and an MS with distinction in management and systems from New York University. She also graduated with a BA from the University of Pennsylvania, majoring in economics (*summa cum laude*).

I want to thank my best friend, Carol Cox, who has helped me to write and relax when needed.

About the reviewers

Ujjwal Karn is a senior software engineer working in the Generative AI space, focusing on enhancing the safety and reliability of large language models. With over a decade of industry and research experience in machine learning, he has developed a unique proficiency in this domain. His research interests span a range of topics, from developing more accurate and efficient language models to investigating new applications for computer vision in real-world settings. As a core contributor to Llama 3 research, Ujjwal continues to drive innovation and progress in the field of AI, and his work has been instrumental in advancing the state of the art in large language model safety.

Aneesh Gadhwal is a senior algorithm developer at Titan Company Limited, working on developing the healthcare ecosystem and focusing on algorithm development for human fitness tracking, vital sign tracking, and health assessment. He graduated from IIT (BHU) in 2021 with a dual degree (B.Tech and M.Tech) in biomedical/medical engineering, where he learned and applied various machine learning and deep learning techniques to biomedical problems. He secured a bronze medal at the BETiC Innovation Challenge in 2018.

He is passionate about using the power of machine learning and deep learning algorithms to improve human health and well-being, and he is always eager to learn new skills and explore new domains in this field.

Table of Contents

Part 2: Mastering LLM Development

3

The Mechanics of Training LLMs   53

4

Advanced Training Strategies   75

5

Fine-Tuning LLMs for Specific Applications   105

6

Testing and Evaluating LLMs   135

Part 3: Deployment and Enhancing LLM Performance

7

Deploying LLMs in Production   161

8

Strategies for Integrating LLMs   185

9

Optimization Techniques for Performance   209

Part 4: Issues, Practical Insights, and Preparing for the Future

11

12

Case Studies – Business Applications and ROI   271

13

The Ecosystem of LLM Tools and Frameworks   293

14

Preparing for GPT-5 and Beyond   317

15

Conclusion and Looking Forward   341

Index   357

Other Books You May Enjoy   372

Preface

In *Decoding Large Language Models*, you will embark on a comprehensive journey, starting with the historical evolution of **Natural Language Processing (NLP)** and the development of **Large Language Models (LLMs)**. The book explores the complex architecture of these models, making intricate concepts such as transformers and attention mechanisms accessible. As the journey progresses, it transitions into the practicalities of training and fine-tuning LLMs, providing hands-on guidance for real-world applications. The narrative then explores advanced optimization techniques and addresses the crucial aspect of ethical considerations in AI. In its final stages, the book offers a forward-looking perspective, preparing you for future developments such as GPT-5. This journey not only educates but also empowers you to skillfully implement and deploy LLMs in various domains.

By the end of this book, you will have gained a thorough understanding of the historical evolution and current state of LLMs in NLP. You will be proficient in the complex architecture of these models, including transformers and attention mechanisms. Your skills will extend to effectively training and fine-tuning LLMs for a variety of real-world applications. You will also have a strong grasp of advanced optimization techniques to enhance model performance. You will be well-versed in the ethical considerations surrounding AI, enabling you to deploy LLMs responsibly. Lastly, you will be prepared for emerging trends and future advancements in the field, such as GPT-5, equipping you to stay at the forefront of AI technology and its applications.

Who this book is for

If you are a technical leader working in NLP, an AI researcher, or a software developer interested in building AI-powered applications, this book is the essential guide to mastering LLMs.

What this book covers

Chapter 1, LLM Architecture, introduces you to the complex anatomy of LLMs. The chapter breaks down the architecture into understandable segments, focusing on the cutting-edge transformer models and the pivotal attention mechanisms they use. A side-by-side analysis with previous RNN models allows you to appreciate the evolution and advantages of current architectures, laying the groundwork for deeper technical understanding.

Chapter 2, How LLMs Make Decisions, provides an in-depth exploration of the decision-making mechanisms in LLMs. It starts by examining how LLMs utilize probability and statistical analysis to process information and predict outcomes. Then, the chapter focuses on the intricate process through which LLMs interpret input and generate responses. Following this, the chapter discusses the various challenges and limitations currently faced by LLMs, including issues of bias and reliability. The chapter concludes by looking at the evolving landscape of LLM decision-making, highlighting advanced techniques and future directions in this rapidly advancing field.

Chapter 3, The Mechanics of Training LLMs, guides you through the intricate process of training LLMs, starting with the crucial task of data preparation and management. The chapter further explores the establishment of a robust training environment, delving into the science of hyperparameter tuning and elaborating on how to address overfitting, underfitting, and other common training challenges, giving you a thorough grounding in creating effective LLMs.

Chapter 4, Advanced Training Strategies, provides more sophisticated training strategies that can significantly enhance the performance of LLMs. It covers the nuances of transfer learning, the strategic advantages of curriculum learning, and the future-focused approaches to multitasking and continual learning. Each concept is solidified with a case study, providing real-world context and applications.

Chapter 5, Fine-Tuning LLMs for Specific Applications, teaches you the fine-tuning techniques tailored to a variety of NLP tasks. From the intricacies of conversational AI to the precision required for language translation and the subtleties of sentiment analysis, you will learn how to customize LLMs for nuanced language comprehension and interaction, equipping you with the skills to meet specific application needs.

Chapter 6, Testing and Evaluating LLMs, explores the crucial phase of testing and evaluating LLMs. This chapter not only covers the quantitative metrics that gauge performance but also stresses the qualitative aspects, including human-in-the-loop evaluation methods. It emphasizes the necessity of ethical considerations and the methodologies for bias detection and mitigation, ensuring that LLMs are both effective and equitable.

Chapter 7, Deploying LLMs in Production, addresses the real-world application of LLMs. You will learn about the strategic deployment of these models, including tackling scalability and infrastructure concerns, ensuring robust security practices, and the crucial role of ongoing monitoring and maintenance to ensure that deployed models remain reliable and efficient.

Chapter 8, Strategies for Integrating LLMs, offers an insightful overview of integrating LLMs into existing systems. It covers the evaluation of LLM compatibility with current technologies, followed by strategies for their seamless integration. The chapter also delves into the customization of LLMs to meet specific system needs, and it concludes with a critical discussion on ensuring security and privacy during the integration process. This concise guide provides essential knowledge to effectively incorporate LLM technology into established systems while maintaining data integrity and system security.

Chapter 9, Optimization Techniques for Performance, introduces advanced techniques that improve the performance of LLMs without sacrificing efficiency. Techniques such as quantization and pruning are discussed in depth, along with knowledge distillation strategies. A focused case study on mobile deployment gives you practical insights into applying these optimizations.

Chapter 10, Advanced Optimization and Efficiency, dives deeper into the technical aspects of enhancing LLM performance. You will explore state-of-the-art hardware acceleration and learn how to manage data storage and representation for optimal efficiency. The chapter provides a balanced view of the trade-offs between cost and performance, a key consideration to deploy LLMs at scale.

Chapter 11, LLM Vulnerabilities, Biases, and Legal Implications, explores the complexities surrounding LLMs, focusing on their vulnerabilities and biases. It discusses the impact of these issues on LLM functionality and the efforts needed to mitigate them. Additionally, the chapter provides an overview of the legal and regulatory frameworks governing LLMs, highlighting intellectual property concerns and the evolving global regulations. It aims to balance the perspectives on technological advancement and ethical responsibilities in the field of LLMs, emphasizing the importance of innovation aligned with regulatory caution.

Chapter 12, Case Studies – Business Applications and ROI, examines the application and **return on investment** (**ROI**) of LLMs in business. It starts with their role in enhancing customer service, showcasing examples of improved efficiency and interaction. The focus then shifts to marketing, exploring how LLMs optimize strategies and content. The chapter then covers LLMs in operational efficiency, particularly in automation and data analysis. It concludes by assessing the ROI from LLM implementations, considering both the financial and operational benefits. Throughout these sections, the chapter presents a comprehensive overview of LLMs' practical business uses and their measurable impacts.

Chapter 13, The Ecosystem of LLM Tools and Frameworks, explores the rich ecosystem of tools and frameworks available for LLMs. It offers a roadmap to navigate the various open source and proprietary tools and comprehensively discusses how to integrate LLMs within existing tech stacks. The strategic role of cloud services in supporting NLP initiatives is also unpacked.

Chapter 14, Preparing for GPT-5 and Beyond, prepares you for the arrival of GPT-5 and subsequent models. It covers the expected features, infrastructure needs, and skillset preparations. The chapter also challenges you to think strategically about potential breakthroughs and how to stay ahead of the curve in a rapidly advancing field.

Chapter 15, Conclusion and Looking Forward, synthesizes the key insights gained throughout the reading journey. It offers a forward-looking perspective on the trajectory of LLMs, pointing you toward resources for continued education and adaptation in the evolving landscape of AI and NLP. The final note encourages you to embrace the LLM revolution with an informed and strategic mindset.

To get the most out of this book

To effectively engage with *Decoding Large Language Models*, you should come equipped with a foundational understanding of machine learning principles, proficiency in a programming language such as Python, a grasp of essential mathematics such as algebra and statistics, and familiarity with NLP basics.

Conventions used

Here are the text conventions used throughout this book.

`Code in text`: Indicates code words in text, database table names, folder names, filenames, file extensions, pathnames, dummy URLs, user input, and Twitter handles. Here is an example: "This contains two basic functions: `add()` and `subtract()`."

A block of code is set as follows:

```
def add(a, b):
    return a + b

def subtract(a, b):
    return a - b
```

Bold: Indicates a new term, an important word, or words that you see on screen. For instance, words in menus or dialog boxes appear in **bold**. Here is an example: "This process, known as **unsupervised learning**, does not require labeled data but instead relies on the patterns inherent in the text itself."

> **Tips or important notes**
> Appear like this.

Get in touch

Feedback from our readers is always welcome.

General feedback: If you have questions about any aspect of this book, email us at `customercare@packtpub.com` and mention the book title in the subject of your message.

Errata: Although we have taken every care to ensure the accuracy of our content, mistakes do happen. If you have found a mistake in this book, we would be grateful if you would report this to us. Please visit `www.packtpub.com/support/errata` and fill in the form.

Piracy: If you come across any illegal copies of our works in any form on the internet, we would be grateful if you would provide us with the location address or website name. Please contact us at copyright@packt.com with a link to the material.

If you are interested in becoming an author: If there is a topic that you have expertise in and you are interested in either writing or contributing to a book, please visit authors.packtpub.com.

Share Your Thoughts

Once you've read *Decoding Large Language Models*, we'd love to hear your thoughts! Scan the QR code below to go straight to the Amazon review page for this book and share your feedback.

https://packt.link/r/1-835-08465-6

Your review is important to us and the tech community and will help us make sure we're delivering excellent quality content.

Download a free PDF copy of this book

Thanks for purchasing this book!

Do you like to read on the go but are unable to carry your print books everywhere?

Is your eBook purchase not compatible with the device of your choice?

Don't worry, now with every Packt book you get a DRM-free PDF version of that book at no cost.

Read anywhere, any place, on any device. Search, copy, and paste code from your favorite technical books directly into your application.

The perks don't stop there, you can get exclusive access to discounts, newsletters, and great free content in your inbox daily

Follow these simple steps to get the benefits:

1. Scan the QR code or visit the link below

https://packt.link/free-ebook/978-1-83508-465-6

2. Submit your proof of purchase
3. That's it! We'll send your free PDF and other benefits to your email directly

Part 1:
The Foundations of Large Language Models (LLMs)

This part provides you with an introduction to LLM architecture, including the anatomy of a language model, transformers and attention mechanisms, **Recurrent Neural Networks** (**RNNs**) and their limitations, and a comparative analysis between transformer and RNN models. It also explains decision making in LLMs, LLM response generation, challenges and limitations in LLM decision making, and advanced techniques and future directions.

This part contains the following chapters:

- *Chapter 1*, *LLM Architecture*
- *Chapter 2*, *How LLMs Make Decisions*

1

LLM Architecture

In this chapter, you'll be introduced to the complex anatomy of **large language models (LLMs)**. We'll break the LLM architecture into understandable segments, focusing on the cutting-edge Transformer models and the pivotal attention mechanisms they use. A side-by-side analysis with previous RNN models will allow you to appreciate the evolution and advantages of current architectures, laying the groundwork for deeper technical understanding.

In this chapter, we're going to cover the following main topics:

- The anatomy of a language model
- Transformers and attention mechanisms
- **Recurrent neural networks (RNNs)** and their limitations
- Comparative analysis – Transformer versus RNN models

By the end of this chapter, you should be able to understand the intricate structure of LLMs, centering on the advanced Transformer models and their key attention mechanisms. You'll also be able to grasp the improvements of modern architectures over older RNN models, which sets the stage for a more profound technical comprehension of these systems.

The anatomy of a language model

In the pursuit of AI that mirrors the depth and versatility of human communication, language models such as GPT-4 emerge as paragons of computational linguistics. The foundation of such a model is its training data – a colossal repository of text drawn from literature, digital media, and myriad other sources. This data is not only vast in quantity but also rich in variety, encompassing a spectrum of topics, styles, and languages to ensure a comprehensive understanding of human language.

The anatomy of a language model such as GPT-4 is a testament to the intersection of complex technology and linguistic sophistication. Each component, from training data to user interaction, works in concert to create a model that not only simulates human language but also enriches the way we interact with machines. It is through this intricate structure that language models hold the promise of bridging the communicative divide between humans and **artificial intelligence (AI)**.

A language model such as GPT-4 operates on several complex layers and components, each serving a unique function to understand, generate, and refine text. Let's go through a comprehensive breakdown.

Training data

The training data for a language model such as GPT-4 is the bedrock upon which its ability to understand and generate human language is built. This data is carefully curated to span an extensive range of human knowledge and expression. Let's discuss the key factors to consider when training data.

Scope and diversity

As an example, the training dataset for GPT-4 is composed of a vast corpus of text that's meticulously selected to cover as broad a spectrum of human language as possible. This includes the following aspects:

- **Literary works**: Novels, poetry, plays, and various forms of narrative and non-narrative literature contribute to the model's understanding of complex language structures, storytelling, and creative uses of language.

- **Informational texts**: Encyclopedias, journals, research papers, and educational materials provide the model with factual and technical knowledge across disciplines such as science, history, arts, and humanities.

- **Web content**: Websites offer a wide range of content, including blogs, news articles, forums, and user-generated content. This helps the model learn current colloquial language and slang, as well as regional dialects and informal communication styles.

- **Multilingual sources**: To be proficient in multiple languages, the training data includes text in various languages, contributing to the model's ability to translate and understand non-English text.

- **Cultural variance**: Texts from different cultures and regions enrich the model's dataset with cultural nuances and societal norms.

Quality and curation

The quality of the training data is crucial. It must have the following attributes:

- **Clean**: The data should be free from errors, such as incorrect grammar or misspellings, unless these are intentional and representative of certain language uses.

- **Accurate**: Accuracy is paramount. Data must be correct and reflect true information to ensure the reliability of the AI's outputs.

- **Varied**: The inclusion of diverse writing styles, from formal to conversational tones, ensures that the model can adapt its responses to fit different contexts.

- **Balanced**: No single genre or source should dominate the training dataset to prevent biases in language generation.

- **Representative**: The data must represent the myriad ways language is used across different domains and demographics to avoid skewed understandings of language patterns.

Training process

The actual training involves feeding textual data into the model, which then learns to predict the next word in a sequence given the words that come before it. This process, known as **supervised learning**, doesn't require labeled data but instead relies on the patterns inherent in the text itself.

Challenges and solutions

The challenges and solutions concerning the training process are as follows:

- **Bias**: Language models can inadvertently learn and perpetuate biases present in training data. To counter this, datasets are often audited for bias, and efforts are made to include a balanced representation.

- **Misinformation**: Texts containing factual inaccuracies can lead to the model learning incorrect information. Curators aim to include reliable sources and may use filtering techniques to minimize the inclusion of misinformation.

- **Updating knowledge**: As language evolves and new information emerges, the training dataset must be updated. This may involve adding recent texts or using techniques to allow the model to learn from new data continuously.

The training data for GPT-4 is a cornerstone that underpins its linguistic capabilities. It's a reflection of human knowledge and language diversity, enabling the model to perform a wide range of language-related tasks with remarkable fluency. The ongoing process of curating, balancing, and updating this data is as critical as the development of the model's architecture itself, ensuring that the language model remains a dynamic and accurate tool for understanding and generating human language.

Tokenization

Tokenization is a fundamental pre-processing step in the training of language models such as GPT-4, serving as a bridge between raw text and the numerical algorithms that underpin **machine learning** (**ML**). Tokenization is a crucial preprocessing step in training language models. It influences the model's ability to understand the text and affects the overall performance of language-related tasks. As models such as GPT-4 are trained on increasingly diverse and complex datasets, the strategies for tokenization continue to evolve, aiming to maximize efficiency and accuracy in representing human language. Here's some in-depth information on tokenization:

- **Understanding tokenization**: Tokenization is the process of converting a sequence of characters into a sequence of tokens, which can be thought of as the building blocks of text. A token is a string of contiguous characters, bounded by spaces or punctuation, that are treated as a group. In language modeling, tokens are often words, but they can also be parts of words (such as subwords or morphemes), punctuation marks, or even whole sentences.

- **The role of tokens**: Tokens are the smallest units that carry meaning in a text. In computational terms, they are the atomic elements that a language model uses to understand and generate language. Each token is associated with a vector in the model, which captures semantic and syntactic information about the token in a high-dimensional space.

- **Tokenization**:

  - **Word-level tokenization**: This is the simplest form and is where the text is split into tokens based on spaces and punctuation. Each word becomes a token.

  - **Subword tokenization**: To address the challenges of word-level tokenization, such as handling unknown words, language models often use subword tokenization. This involves breaking down words into smaller meaningful units (subwords), which helps the model generalize better to new words. This is particularly useful for handling inflectional languages, where the same root word can have many variations.

  - **Byte-pair encoding (BPE)**: BPE is a common subword tokenization method. It starts with a large corpus of text and combines the most frequently occurring character pairs iteratively. This continues until a vocabulary of subword units is built that optimizes for the corpus's most common patterns.

- **SentencePiece**: SentencePiece is a tokenization algorithm that doesn't rely on predefined word boundaries and can work directly on raw text. This means it processes the text in its raw form without needing prior segmentation into words. This method is different from approaches such as BPE, which often require initial text segmentation. Working directly on raw text allows SentencePiece to be language-agnostic, making it particularly effective for languages that don't use whitespace to separate words, such as Japanese or Chinese. In contrast, BPE typically works on pre-tokenized text, where words are already separated, which might limit its effectiveness for certain languages without explicit word boundaries.

  By not depending on pre-defined boundaries, SentencePiece can handle a wider variety of languages and scripts, providing a more flexible and robust tokenization method for diverse linguistic contexts.

The process of tokenization

The process of tokenization in the context of language models involves several steps:

1. **Segmentation**: Splitting the text into tokens based on predefined rules or learned patterns.

2. **Normalization**: Sometimes, tokens are normalized to a standard form. For instance, 'USA' and 'U.S.A.' might be normalized to a single form.

3. **Vocabulary indexing**: Each unique token is associated with an index in a vocabulary list. The model will use these indices, not the text itself, to process the language.

4. **Vector representation**: Tokens are converted into numerical representations, often as one-hot vectors or embeddings, which are then fed into the model.

The importance of tokenization

Tokenization plays a critical role in the performance of language models by supporting the following aspects:

- **Efficiency**: It enables the model to process large amounts of text efficiently by reducing the size of the vocabulary it needs to handle.

- **Handling unknown words**: By breaking words into subword units, the model can handle words it hasn't seen before, which is particularly important for open domain models that encounter diverse text.

- **Language flexibility**: Subword and character-level tokenization enable the model to work with multiple languages more effectively than word-level tokenization. This is because subword and character-level approaches break down text into smaller units, which can capture commonalities between languages and handle various scripts and structures. For example, many languages share roots, prefixes, and suffixes that can be understood at the subword level. This granularity helps the model generalize better across languages, including those with rich morphology or unique scripts.

- **Semantic and syntactic learning**: Proper tokenization allows the model to learn the relationships between different tokens, capturing the nuances of language.

Challenges of tokenization

The following challenges are associated with tokenization:

- **Ambiguity**: Tokenization can be ambiguous, especially in languages with complex word formations or in the case of homographs (words that are spelled the same but have different meanings)

- **Context dependency**: The meaning of a token can depend on its context, which is not always considered in simple tokenization schemes

- **Cultural differences**: Different cultures may have different tokenization needs, such as compound words in German or lack of spaces in Chinese

Neural network architecture

The neural network architecture of models such as GPT-4 is a sophisticated and intricate system designed to process and generate human language with great proficiency. The Transformer neural architecture, which is the backbone of GPT-4, represents a significant leap in the evolution of neural network designs for language processing.

The Transformer architecture

The Transformer architecture was introduced in a paper titled *Attention Is All You Need*, by Vaswani et al., in 2017. It represents a departure from earlier sequence-to-sequence models that used **recurrent neural network** (**RNN**) or **convolutional neural network** (**CNN**) layers. The Transformer model is designed to handle sequential data without the need for these recurrent structures, thus enabling more parallelization and reducing training times significantly. The Transformer relies entirely on self-attention mechanisms to process data in parallel, which allows for significantly faster computation.

Self-attention mechanisms

An encoder processes input data into a fixed representation for further use by the model, while a decoder transforms the fixed representation back into a desired output format, such as text or sequences. Self-attention, sometimes called intra-attention, is a mechanism that allows each position in the encoder to attend to all positions in the previous layer of the encoder. Similarly, each position in the decoder can attend to all positions in the encoder and all positions up to and including that position in the decoder. This mechanism is vital for the model's ability to understand the context and relationships within the input data.

Self-attention at work

It calculates a set of attention scores for each token in the input data, determining how much focus it should put on other parts of the input when processing a particular token.

These scores are used to create a weighted combination of value vectors, which then becomes the input to the next layer or the output of the model.

Multi-head self-attention

A pivotal aspect of the Transformer's attention mechanism is that it uses multiple "heads," meaning that it runs the attention mechanism several times in parallel. Each "head" learns different aspects of the data, which allows the model to capture various types of dependencies in the input: syntactic, semantic, and positional.

The advantages of multi-head attention are as follows:

- It gives the model the ability to pay attention to different parts of the input sequence differently, which is similar to considering a problem from different perspectives
- Multiple representations of each token are learned, which enriches the model's understanding of each token in its context

Position-wise feedforward networks

After the attention sub-layers in each layer of the encoder and decoder, there's a fully connected feedforward network. This network applies the same linear transformation to each position separately and identically. This part of the model can be seen as a processing step that refines the output of the attention mechanism before passing it on to the next layer.

The function of the feedforward networks is to provide the model with the ability to apply more complex transformations to the data. This part of the model can learn and represent non-linear dependencies in the data, which are crucial for capturing the complexities of language.

Layer normalization and residual connections

The Transformer architecture utilizes layer normalization and residual connections to enhance training stability and enable deeper models to be trained:

- **Layer normalization**: It normalizes the inputs across the features for each token independently and is applied before each sub-layer in the Transformer, enhancing training stability and model performance.

- **Residual connections**: Each sub-layer in the Transformer, be it an attention mechanism or a feedforward network, has a residual connection around it, followed by layer normalization. This means that the output of each sub-layer is added to its input before being passed on, which helps mitigate the vanishing gradients problem, allowing for deeper architectures. The vanishing gradients problem occurs during training deep neural networks when gradients of the loss function diminish exponentially as they're backpropagated through the layers, leading to extremely small weight updates and hindering learning.

The neural network architecture of GPT-4, based on the Transformer, is a testament to the evolution of ML techniques in **natural language processing** (**NLP**). The self-attention mechanisms enable the model to focus on different parts of the input, multi-head attention allows it to capture multiple dependency types, and the position-wise feedforward networks contribute to understanding complex patterns. Layer normalization and residual connections ensure that the model can be trained effectively even when it is very deep. All these components work together in harmony to allow models such as GPT-4 to generate text that is contextually rich, coherent, and often indistinguishable from text written by humans.

Embeddings

In the context of language models such as GPT-4, embeddings are a critical component that enables these models to process and understand text at a mathematical level. Embeddings transform discrete tokens – such as words, subwords, or characters – into continuous vectors, from which a vector operation can be applied to the embeddings. Let's break down the concept of embeddings and their role in language models:

- **Word embeddings**: Word embeddings are the most direct form of embeddings, where each word in the model's vocabulary is transformed into a high-dimensional vector. These vectors are learned during the training process.

Let's take a look at the characteristics of word embeddings:

- **Dense representation**: Each word is represented by a dense vector, typically with several hundred dimensions, as opposed to sparse, high-dimensional representations like one-hot encoding.

- **Semantic similarity**: Semantically similar words tend to have embeddings that are close to each other in the vector space. This allows the model to understand synonyms, analogies, and general semantic relationships.

- **Learned in context**: The embeddings are learned based on the context in which the words appear, so the vector for a word captures not just the word itself but also how it's used.

- **Subword embeddings**: For handling out-of-vocabulary words and morphologically rich languages, subword embeddings break down words into smaller components. This allows the model to generate embeddings for words it has never seen before, based on the subword units.

- **Positional embeddings**: Since the Transformer architecture that's used by GPT-4 doesn't inherently process sequential data in order, positional embeddings are added to give the model information about the position of words in a sequence.

Let's look at the features of positional embeddings:

- **Sequential information**: Positional embeddings encode the order of the tokens in the sequence, allowing the model to distinguish between "John plays the piano" and "The piano plays John," for example.

- **Added to word embeddings**: These positional vectors are typically added to the word embeddings before they're inputted into the Transformer layers, ensuring that the position information is carried through the model.

In understanding the architecture of language models, we must understand two fundamental components:

- **Input layer**: In language models, embeddings form the input layer, transforming tokens into a format that the neural network can work with

- **Training process**: During training, the embeddings are adjusted along with the other parameters of the model to minimize the loss function, thus refining their ability to capture linguistic information

The following are two critical stages in the development and enhancement of language models:

- **Initialization**: Embeddings can be randomly initialized and learned from scratch during training, or they can be pre-trained using unsupervised learning on a large corpus of text and then fine-tuned for specific tasks.

- **Transfer learning**: Embeddings can be transferred between different models or tasks. This is the principle behind models such as BERT, where the embeddings learned from one task can be applied to another.

Challenges and solutions

There are challenges you must overcome when using embeddings. Let's go through them and learn how to tackle them:

- **High dimensionality**: Embeddings are highly dimensional, which can make them computationally expensive. Dimensionality reduction techniques and efficient training methods can be employed to manage this.

- **Context dependence**: A word might have different meanings in different contexts. Models such as GPT-4 use the surrounding context to adjust the embeddings during the self-attention phase, addressing this challenge.

In summary, embeddings are a foundational element of modern language models, transforming the raw material of text into a rich, nuanced mathematical form that the model can learn from. By capturing semantic meaning and encoding positional information, embeddings allow models such as GPT-4 to generate and understand language with a remarkable degree of sophistication.

Transformers and attention mechanisms

Attention mechanisms in language models such as GPT-4 are a transformative innovation that enables the model to selectively focus on specific parts of the input data, much like how human attention allows us to concentrate on particular aspects of what we're reading or listening to. Here's an in-depth explanation of how attention mechanisms function within these models:

- **Concept of attention mechanisms**: The term "attention" in the context of neural networks draws inspiration from the attentive processes observed in human cognition. The attention mechanism in neural networks was introduced to improve the performance of encoder-decoder architectures, especially in tasks such as machine translation, where the model needs to correlate segments of the input sequence with the output sequence.

- **Functionality of attention mechanisms**:

  - **Contextual relevance**: Attention mechanisms weigh the elements of the input sequence based on their relevance to each part of the output. This allows the model to create a context-sensitive representation of each word when making predictions.

  - **Dynamic weighting**: Unlike previous models, which treated all parts of the input sequence equally or relied on fixed positional encoding, attention mechanisms dynamically assign weights to different parts of the input for each output element.

Types of attention

The following types of attention exist in neural networks:

- **Global attention**: The model considers all the input tokens for each output token.

- **Local attention**: The model only focuses on a subset of input tokens that are most relevant to the current output token.

- **Self-attention**: In this scenario, the model attends to all positions within a single sequence, allowing each position to be informed by the entire sequence. This type is used in the Transformer architecture and enables parallel processing of sequences.

- **Multi-head attention**: Multi-head attention is a mechanism in neural networks that allows the model to focus on different parts of the input sequence simultaneously by computing attention scores in parallel across multiple heads.

- **Relative attention**: Relative attention is a mechanism that enhances the attention model by incorporating information about the relative positions of tokens, allowing the model to consider the positional relationships between tokens more effectively.

The process of attention in Transformers

In the case of the Transformer model, the attention process involves the following steps:

1. **Attention scores**: The model computes scores to determine how much attention to pay to other tokens in the sequence for each token.

2. **Scaled dot-product attention**: This specific type of attention that's used in Transformers calculates the scores by taking the dot product of the query with all keys, dividing each by the square root of the dimensionality of the keys (to achieve more stable gradients), and then applying a softmax function to obtain the weights for the values.

3. **Query, key, and value vectors**: Every token is associated with three vectors – a query vector, a key vector, and a value vector. The attention scores are calculated using the query and key vectors, and these scores are used to weigh the value vectors.

4. **Output sequence**: The weighted sum of the value vectors, informed by the attention scores, becomes the output for the current token.

Advancements in language model capabilities, such as the following, have significantly contributed to the refinement of NLP technologies:

- **Handling long-range dependencies**: They allow the model to handle long-range dependencies in text by focusing on relevant parts of the input, regardless of their position.

- **Improved translation and summarization**: In tasks such as translation, the model can focus on the relevant word or phrase in the input sentence when translating a particular word, leading to more accurate translations.

- **Interpretable model behavior**: Attention maps can be inspected to understand which parts of the input the model is focusing on when making predictions, adding an element of interpretability to these otherwise "black-box" models.

The following facets are crucial considerations in the functionality of attention mechanisms within language models:

- **Computational complexity**: Attention can be computationally intensive, especially with long sequences. Optimizations such as "attention heads" in multi-head attention allow for parallel processing to mitigate this.

- **Contextual comprehension**: While attention allows the model to focus on relevant parts of the input, ensuring that this focus accurately represents complex relationships in the data remains a challenge that requires ongoing refinement of the attention mechanisms.

Attention mechanisms endow language models with the ability to parse and generate text in a context-aware manner, closely mirroring the nuanced capabilities of human language comprehension and production. Their role in the Transformer architecture is pivotal, contributing significantly to the state-of-the-art performance of models such as GPT-4 in a wide range of language processing tasks.

Decoder blocks

Decoder blocks are an essential component in the architecture of many Transformer-based models, although with a language model such as GPT-4, which is used for tasks such as language generation, the architecture is slightly different as it's based on a decoder-only structure. Let's take a detailed look at the functionality and composition of these decoder blocks within the context of GPT-4.

The role of decoder blocks in GPT-4

In traditional Transformer models, such as those used for translation, there are both encoder and decoder blocks – the encoder processes the input text while the decoder generates the translated output. GPT-4, however, uses a slightly modified version of this architecture that consists solely of what can be described as decoder blocks.

These blocks are responsible for generating text and predicting the next token in a sequence given the previous tokens. This is a form of autoregressive generation where the model predicts one token at a time sequentially using the output as part of the input for the next prediction.

The structure of decoder blocks

Each decoder block in GPT-4's architecture is composed of several key components:

- **Self-attention mechanism**: At the core of each decoder block is a self-attention mechanism that allows the block to consider the entire sequence of tokens generated so far. This mechanism is crucial for understanding the context of the sequence up to the current point.

- **Masked attention**: Since GPT-4 generates text autoregressively, it uses masked self-attention in the decoder blocks. This means that when predicting a token, the attention mechanism only considers the previous tokens and not any future tokens, which the model should not have access to.

- **Multi-head attention**: Within the self-attention mechanism, GPT-4 employs multi-head attention. This allows the model to capture different types of relationships in the data – such as syntactic and semantic connections – by processing the sequence in multiple different ways in parallel.

- **Position-wise feedforward networks**: Following the attention mechanism, each block contains a feedforward neural network. This network applies further transformations to the output of the attention mechanism and can capture more complex patterns that attention alone might miss.

- **Normalization and residual connections**: Each sub-layer (both the attention mechanism and the feedforward network) in the decoder block is followed by normalization and includes a residual connection from its input, which helps to prevent the loss of information through the layers and promotes more effective training of deep networks.

Functioning of decoder blocks

The process of generating text with decoder blocks entails the following steps:

1. **Token generation**: Starting with an initial input (such as a prompt), the decoder blocks generate one token at a time.

2. **Context integration**: The self-attention mechanism integrates the context from the entire sequence of generated tokens to inform the prediction of the next token.

3. **Refinement**: The feedforward network refines the output from the attention mechanism, and the result is normalized to ensure that it fits well within the expected range of values.

4. **Iterative process**: This process is repeated iteratively, with each new token being generated based on the sequence of all previous tokens.

The significance of decoder blocks

Decoder blocks in GPT-4 are significant due to the following reasons:

- **Context-awareness**: Decoder blocks allow GPT-4 to generate text that's contextually coherent and relevant, maintaining consistency across long passages of text

- **Complex pattern learning**: The combination of attention mechanisms and feedforward networks enables the model to learn and generate complex patterns in language, from simple syntactic structures to nuanced literary devices

- **Adaptive generation**: The model can adapt its generation strategy based on the input it receives, making it versatile across different styles, genres, and topics

The decoder blocks in GPT-4's architecture are sophisticated units of computation that perform the intricate task of text generation. Through a combination of attention mechanisms and neural networks, these blocks enable the model to produce text that closely mimics human language patterns, with each block building upon the previous ones to generate coherent and contextually rich language.

Parameters

The parameters of a neural network, such as GPT-4, are the elements that the model learns from the training data. These parameters are crucial for the model to make predictions and generate text that's coherent and contextually appropriate.

Let's understand the parameters of neural networks:

- **Definition**: In ML, parameters are the configuration variables that are internal to the model that are learned from the data. They're adjusted through the training process.

- **Weights and biases**: The primary parameters in neural networks are the weights and biases in each neuron. Weights determine the strength of the connection between two neurons, while biases are added to the output of the neuron to shift the activation function.

Certain aspects are pivotal in the development and refinement of advanced language models such as GPT-4:

- **Scale**: GPT-4 is notable for its vast number of parameters. The exact number of parameters is a design choice that affects the model's capacity to learn from data. More parameters generally means a higher capacity for learning complex patterns.

- **Fine-tuning**: The values of these parameters are fine-tuned during the training process to minimize the loss, which is a measure of the difference between the model's predictions and the actual data.

- **Gradient descent**: Parameters are typically adjusted using algorithms such as gradient descent, where the model's loss is calculated, and gradients are computed that indicate how the parameters should be changed to reduce the loss.

The following key factors are central to the sophistication of models such as GPT-4:

- **Capturing linguistic nuances**: Parameters enable the model to capture the nuances of language, including grammar, style, idiomatic expressions, and even the tone of text

- **Contextual understanding**: In GPT-4, parameters help in understanding context, which is crucial for generating text that follows from the given prompt or continues a passage coherently

- **Knowledge representation**: They also allow the model to "remember" factual information it has learned during training, enabling it to answer questions or provide factually accurate explanations

The following optimization techniques are essential in the iterative training process of neural networks:

- **Backpropagation**: During training, the model uses a backpropagation algorithm to adjust the parameters. The model makes a prediction, calculates the error, and then propagates this error back through the network to update the parameters.

- **Learning rate**: The learning rate is a hyperparameter that determines the size of the steps taken during gradient descent. It's crucial for efficient training as too large a rate can cause overshooting and too small a rate can cause slow convergence.

The following challenges are critical considerations:

- **Overfitting**: With more parameters, there's a risk that the model will overfit to the training data, capturing noise rather than the underlying patterns

- **Computational resources**: Training models with a vast number of parameters requires significant computational resources, both in terms of processing power and memory

- **Environmental impact**: The energy consumption for training such large models has raised concerns about the environmental impact of AI research

Parameters are the core components of GPT-4 that enable it to perform complex tasks such as language generation. They are the key to the model's learning capabilities, allowing it to absorb a wealth of information from the training data and apply it when generating new text. The vast number of parameters in GPT-4 allows for an unparalleled depth and breadth of knowledge representation, contributing to its state-of-the-art performance in a wide range of language processing tasks. However, the management of these parameters poses significant technical and ethical challenges that continue to be an active area of research and discussion in the field of AI.

Fine-tuning

Fine-tuning is a critical process in ML, especially in the context of sophisticated models such as GPT-4. It involves taking a pre-trained model and continuing the training process with a smaller, more specialized dataset to adapt the model to specific tasks or improve its performance on certain types of text. This stage is pivotal for tailoring a general-purpose model to specialized applications. Let's take a closer look at the process and the importance of fine-tuning.

The process of fine-tuning

The fine-tuning process comprises the following steps:

1. **Initial model training**: First, GPT-4 is trained on a vast, diverse dataset so that it can learn a wide array of language patterns and information. This is known as supervised pre-training.

2. **Selecting a specialized dataset**: For fine-tuning, a dataset is chosen that closely matches the target task or domain. This dataset is usually much smaller than the one used for initial training and is often labeled, providing clear examples of the desired output.

3. **Continued training**: The model is then further trained (fine-tuned) on this new dataset. The pre-trained weights are adjusted to better suit the specifics of the new data and tasks.

4. **Task-specific adjustments**: During fine-tuning, the model may also undergo architectural adjustments, such as adding or modifying output layers, to better align with the requirements of the specific task.

The importance of fine-tuning

Let's review a few aspects of fine-tuning that are important:

*   **Improved performance**: Fine-tuning allows the model to significantly improve its performance on tasks such as sentiment analysis, question-answering, or legal document analysis by learning from task-specific examples

*   **Domain adaptation**: It helps the model to adapt to the language and knowledge of a specific domain, such as medical or financial texts, where understanding specialized vocabulary and concepts is crucial

*   **Customization**: For businesses and developers, fine-tuning offers a way to customize the model to their specific needs, which can greatly enhance the relevance and utility of the model's outputs

Techniques in fine-tuning

When it comes to working with fine-tuning, some techniques must be implemented:

*   **Transfer learning**: Fine-tuning is a form of transfer learning where knowledge gained while solving one problem is applied to a different but related problem.

*   **Learning rate**: The learning rate during fine-tuning is usually smaller than during initial training, allowing for subtle adjustments to the model's weights without overwriting what it has already learned.

*   **Regularization**: Techniques such as dropout or weight decay might be adjusted during fine-tuning to prevent overfitting to the smaller dataset.

*   **Quantization**: Quantization is the process of reducing the precision of the numerical values in a model's parameters and activations, often from floating-point to lower bit-width integers, to decrease memory usage and increase computational efficiency.

*   **Pruning**: Pruning is a technique that involves removing less important neurons or weights from a neural network to reduce its size and complexity, thereby improving efficiency and potentially mitigating overfitting. Overfitting happens when a model learns too much from the training data, including its random quirks, making it perform poorly on new, unseen data.

*   **Knowledge distillation**: Knowledge distillation is a technique where a smaller, simpler model is trained to replicate the behavior of a larger, more complex model, effectively transferring knowledge from the "teacher" model to the "student" model.

Challenges in fine-tuning

Fine-tuning also has its own set of challenges:

- **Data quality**: The quality of the fine-tuning dataset is paramount. Poor quality or non-representative data can lead to model bias or poor generalization.

- **Balancing specificity with general knowledge**: There is a risk of overfitting to the fine-tuning data, which can cause the model to lose some of its general language abilities.

- **Resource intensity**: While less resource-intensive than the initial training, fine-tuning still requires substantial computational resources, especially when done repeatedly or for multiple tasks.

- **Adversarial attacks**: Adversarial attacks involve deliberately modifying inputs to an ML model in a way that causes the model to make incorrect predictions or classifications. They're conducted to expose vulnerabilities in ML models, test their robustness, and improve security measures by understanding how models can be deceived.

Applications of fine-tuned models

Fine-tuned models can be implemented in different areas:

- **Personalized applications**: Fine-tuned models can provide personalized experiences in applications such as chatbots, where the model can be adapted to the language and preferences of specific user groups

- **Compliance and privacy**: For sensitive applications, fine-tuning can ensure that a model complies with specific regulations or privacy requirements by training on appropriate data

- **Language and locale specificity**: Fine-tuning can adapt models so that they understand and generate text in specific dialects or regional languages, making them more accessible and user-friendly for non-standard varieties of language

In summary, fine-tuning is a powerful technique for enhancing the capabilities of language models such as GPT-4, enabling them to excel in specific tasks and domains. By leveraging the broad knowledge learned during initial training and refining it with targeted data, fine-tuning bridges the gap between general-purpose language understanding and specialized application requirements.

Outputs

The output generation process in a language model such as GPT-4 is a complex sequence of steps that results in the creation of human-like text. This process is built on the foundation of predicting the next token in a sequence. Here's a detailed exploration of how GPT-4 generates outputs.

- **Token probability calculation**:
  - **Probabilistic model**: GPT-4, at its core, is a probabilistic model. For each token it generates, it calculates a distribution of probabilities over all tokens in its vocabulary, which can include tens of thousands of different tokens.

- **Softmax function**: The model uses a softmax function on the logits (the raw predictions of the model) to create this probability distribution. The softmax function exponentiates and normalizes the logits, ensuring that the probabilities sum up to one.

- **Token selection**:

  - **Highest probability**: Once the probabilities are calculated, the model selects the token with the highest probability as the next piece of output. This is known as greedy decoding. However, this isn't the only method available for selecting the next token.

  - **Sampling methods**: To introduce variety and handle uncertainty, the model can also use different sampling methods. For instance, "top-k sampling" limits the choice to the k most likely next tokens, while "nucleus sampling" (top-p sampling) chooses from a subset of tokens that cumulatively make up a certain probability.

- **Autoregressive generation**:

  - **Sequential process**: GPT-4 generates text autoregressively, meaning that it generates one token at a time, and each token is conditioned on the previous tokens in the sequence. After generating a token, it's added to the sequence, and the process is repeated.

  - **Context update**: With each new token generated, the model updates its internal representation of the context, which influences the prediction of subsequent tokens.

- **Stopping criteria**:

  - **End-of-sequence token**: The model is typically programmed to recognize a special token that signifies the end of a sequence. When it predicts this token, the output generation process stops.

  - **Maximum length**: Alternatively, the generation can be stopped after it reaches a maximum length to prevent overly verbose outputs or when the model starts to loop or diverge semantically.

- **Refining outputs**:

  - **Beam search**: Instead of selecting the single best next token at each step, beam search explores several possible sequences simultaneously, keeping a fixed number of the most probable sequences (the "beam width") at each time step

  - **Human-in-the-loop**: In some applications, outputs may be refined with human intervention, where a user can edit or guide the model's generation

- **Challenges in output generation**:

  - **Maintaining coherence**: Ensuring that the output remains coherent over longer stretches of text is a significant challenge, especially as the context the model must consider grows

- **Avoiding repetition**: Language models can sometimes fall into repetitive loops, particularly with greedy decoding

- **Handling ambiguity**: Deciding on the best output when multiple tokens seem equally probable can be difficult, and different sampling strategies may be employed to address this

- **Generating diverse and creative outputs**: Producing varied and imaginative responses while avoiding bland or overly generic text is crucial for creating engaging and innovative content

- **Applications of the output generation process**:

  - **Conversational AI**: Generating outputs that can engage in dialog with users

  - **Content creation**: Assisting in writing tasks by generating articles, stories, or code

  - **Language translation**: Translating text from one language into another by generating text in the target language

The output generation of GPT-4 is a sophisticated interplay of probability calculation, sampling strategies, and sequence building. The model's ability to generate coherent and contextually appropriate text hinges on its complex internal mechanisms, which allow it to approximate the intricacy of human language. These outputs are not just a simple prediction of the next word but the result of a highly dynamic and context-aware process.

Applications

Language models such as GPT-4, with their advanced capabilities in understanding and generating human-like text, are applied across a wide array of domains, revolutionizing the way we interact with technology and handle information. Here's an in-depth look at various applications where language models have a significant impact:

- **Text completion and autocorrection**:

  - **Writing assistance**: Language models offer suggestions to complete sentences or paragraphs, helping writers to express ideas more efficiently

  - **Email and messaging**: They can predict what a user intends to type next, improving speed and accuracy in communication

- **Translation**:

  - **Machine translation**: These models can translate text between languages, making global communication more accessible

  - **Real-time interpretation**: They enable real-time translation services for speech-to-text applications, breaking down language barriers in conversations

- **Summarization**:

  - **Information condensation**: Language models can distill long articles, reports, or documents into concise summaries, saving time and making information consumption more manageable

  - **Customized digests**: They can create personalized summaries of content based on user interests or queries

- **Question answering**:

  - **Information retrieval**: Language models can answer queries by understanding and sourcing information from large databases or the internet

  - **Educational tools**: They assist in educational platforms, providing students with explanations and helping with homework

- **Content generation**:

  - **Creative writing**: They can assist in generating creative content such as poetry, stories, or even music lyrics

  - **Marketing and copywriting**: Language models are used to generate product descriptions, advertising copy, and social media posts

- **Sentiment analysis**:

  - **Market research**: By analyzing customer feedback, reviews, and social media mentions, language models can gauge public sentiment toward products, services, or brands

  - **Crisis management**: They help organizations monitor and respond to public sentiment in times of crisis or controversy

- **Personal assistants**:

  - **Virtual assistants**: Language models power virtual assistants in smartphones, home devices, and customer service chatbots, enabling them to understand and respond to user requests

  - **Accessibility**: They support the creation of tools that assist individuals with disabilities by generating real-time descriptive text for visual content or interpreting sign language

- **Code generation and automation**:

  - **Software development**: They assist in generating code snippets, debugging, or even creating simple programs, increasing developer productivity

  - **Automation of repetitive tasks**: Language models can automate routine documentation or reporting tasks, freeing up human resources for more complex activities

- **Fine-tuning for specialized tasks**:

  - **Legal and medical fields**: Language models can be fine-tuned to understand jargon and generate documents specific to these fields

  - **Scientific research**: They can summarize research papers, suggest potential areas of study, or even generate hypotheses based on existing data

- **Language learning**:

  - **Educational platforms**: Language models support language learning platforms by providing conversation practice and grammar correction

  - **Cultural exchange**: They facilitate the understanding of different cultures by providing insights into colloquial and idiomatic expressions

- **Ethical and creative writing**:

  - **Bias detection**: They can be used to detect and correct biases in writing, promoting more ethical and inclusive content creation

  - **Storytelling**: Language models contribute to interactive storytelling experiences, adapting narratives based on user input or actions

The applications of language models such as GPT-4 are diverse and continually expanding as technology advances. They have become integral tools in fields ranging from communication to education, content creation, and beyond, offering significant benefits in terms of efficiency, accessibility, and the democratization of information. As these models become more sophisticated, their integration into daily tasks and specialized industries is poised to become even more seamless and impactful.

Ethical considerations

The deployment and development of language models such as GPT-4 raise several ethical considerations that must be addressed by developers, policymakers, and society as a whole. These considerations encompass a range of issues, from the inherent biases in training data to the potential for spreading misinformation and the socioeconomic impacts. Here's a detailed examination of these concerns:

- **Bias in language models**:

  - **Training data**: Language models learn from existing text data, which can contain historical and societal biases. These biases can be reflected in the model's outputs, perpetuating stereotypes or unfair portrayals of individuals or groups.

  - **Representation**: The data used to train these models may not equally represent different demographics, leading to outputs that are less accurate or relevant for underrepresented groups.

- **Misinformation and deception**:

  - **Spread of misinformation**: If not carefully monitored, language models can generate plausible-sounding but inaccurate or misleading information, contributing to the spread of misinformation

  - **Manipulation and deception**: There's a risk of these models being used to create fake news, impersonate individuals, or generate deceptive content, which can have serious societal consequences

- **Impact on jobs**:

  - **Automation**: As language models take over tasks traditionally performed by humans, such as writing reports or answering customer service queries, there can be an impact on employment in those sectors

  - **Skill displacement**: Workers may need to adapt and develop new skills as their roles evolve with the integration of AI technologies

  - **Copyright and intellectual property rights**: The use of AI-generated content raises concerns about determining ownership and protecting creative works

- **Privacy**:

  - **Data usage**: The data used to train language models can contain sensitive personal information. Ensuring that this data is used responsibly and that individuals' privacy is protected is a significant concern.

  - **Consent**: In many cases, the individuals whose data is used to train these models may not have given explicit consent for their information to be used in this way.

- **Transparency and accountability**:

  - **Understanding model decisions**: It can be challenging to understand how language models come to certain conclusions or decisions, leading to calls for greater transparency

  - **Accountability**: When a language model produces a harmful output, determining who is responsible – the developer, the user, or the model itself – can be complex

- **Human interaction**:

  - **Dependency**: There's a concern that over-reliance on language models could diminish human critical thinking and interpersonal communication skills

  - **Human-AI relationship**: How humans interact with AI, and the trust they place in automated systems, are ethical considerations, particularly when these systems mimic human behavior

- **Mitigating ethical risks**:
  - **Bias monitoring and correction**: Developers are employing various techniques to detect and mitigate biases in models, including diversifying training data and adjusting model parameters
  - **Transparency measures**: Initiatives to make the workings of AI models more understandable and explainable are underway to enhance transparency
  - **Regulation and policy**: Governments and international bodies are beginning to develop regulations and frameworks to ensure ethical AI development and deployment
- **Societal dialog**:
  - **Public discourse**: Engaging the public in a dialog about the role of AI in society and the ethical considerations of language models is crucial for responsible development
  - **Interdisciplinary approach**: Collaboration between technologists, ethicists, sociologists, and other stakeholders is essential to address the multifaceted ethical issues posed by AI

In conclusion, the ethical considerations surrounding language models are multifaceted and require ongoing attention and action. As these models become more integrated into various aspects of society, it's vital to proactively address these issues to ensure that the benefits of AI are distributed fairly and that potential harms are mitigated. The responsible development and deployment of language models necessitate a commitment to ethical principles, transparency, and inclusive dialog.

Safety and moderation

Ensuring the safety and integrity of language models such as GPT-4 is crucial for their responsible use. Safety and moderation mechanisms are designed to prevent the generation of harmful content, which includes anything from biased or offensive language to the dissemination of false information. Let's take an in-depth look at the various strategies and research initiatives that aim to bolster the safety and moderation of these powerful tools:

- **Content filtering**:
  - **Preventative measures**: Language models often incorporate filters that preemptively prevent the generation of content that could be harmful, such as hate speech, explicit language, or violent content
  - **Dynamic filtering**: These systems can be dynamic, using feedback loops to continuously improve the detection and filtering of harmful content based on new data and patterns
- **User input moderation**:
  - **Input scrubbing**: Safety mechanisms can include analyzing and scrubbing user inputs to prevent the model from being prompted to generate unsafe content

- **Contextual understanding**: Moderation tools are being developed to understand the context of queries better, which helps in distinguishing between potentially harmful and benign requests

- **Reinforcement learning from human feedback (RLHF)**:

  - **Iterative training**: By incorporating human feedback into the training loop, language models can learn what types of content are considered unsafe or undesirable over time

  - **Value alignment**: RLHF is part of ensuring the model's outputs align with human values and ethical standards

- **Red teaming**:

  - **Adversarial testing**: Red teams are used to probe and test the model for vulnerabilities, deliberately attempting to make it generate unsafe content to improve defense mechanisms

  - **Continuous evaluation**: This process helps in identifying weaknesses in the model's safety measures, allowing developers to patch and improve them

- **Transparency and explainability**:

  - **Model insights**: Developing ways to explain why a model generates certain outputs is key to building trust and ensuring moderation systems are working correctly

  - **Audit trails**: Keeping records of model interactions can help you track and understand how and why harmful content might slip through, leading to better moderation

- **Collaboration and standards**:

  - **Cross-industry standards**: There's ongoing work to establish industry-wide standards for what constitutes harmful content and how to deal with it

  - **Open research**: Many organizations are engaging in open research collaborations to tackle the challenge of AI safety, sharing insights and breakthroughs

- **Impact monitoring**:

  - **Real-world monitoring**: Deployed models are monitored to see how they interact with users in real-world scenarios, providing data to refine safety mechanisms

  - **Feedback loops**: User reporting tools and feedback mechanisms allow developers to collect data on potential safety issues that arise during use

- **Ethical and cultural sensitivity**:

  - **Global perspectives**: Safety systems are designed to be sensitive to a diverse range of ethical and cultural norms, which can vary widely across different user bases

- **Inclusive design**: By involving a diverse group of people in the design and testing of moderation systems, developers can better ensure that safety measures are inclusive and equitable

Safety and moderation in language models are multifaceted challenges that involve both technological solutions and human oversight. The goal is to create robust systems that can adapt and respond to the complex, evolving landscape of human communication. As language models continue to be integrated into more aspects of society, the importance of these safety mechanisms cannot be overstated. They are vital for ensuring that the benefits of AI can be enjoyed widely while minimizing the risks of harm and misuse. The ongoing research and development in this area are critical to building trust and establishing the sustainable use of AI technologies in our daily lives.

User interaction

User interaction plays a crucial role in the functioning and continuous improvement of language models such as GPT-4. The model's design accommodates and learns from the various ways in which users engage with it, which can include providing prompts, feedback, and corrections. Let's take an in-depth look at the significance of user interaction with language models:

- **Prompt engineering**:

  - **Prompt design**: The way a user crafts a prompt can greatly influence the model's response. Users have learned to use "prompt engineering" or "prompt crafting" to guide the model toward generating the desired output.

  - **Instruction following**: GPT-4 and similar models are designed to follow user instructions as closely as possible, making the clarity and specificity of prompts vital.

  - **Security prospects in user interaction**: Ensuring secure and safe interactions with the model is crucial as inappropriate or harmful prompts can lead to unintended and potentially dangerous outputs.

- **Feedback loops**:

  - **Reinforcement learning**: Some language models use reinforcement learning techniques, where user feedback on the model's outputs can be used as a signal to adjust the model's parameters

  - **Continuous learning**: Though GPT-4 doesn't learn from interactions after its initial training period due to fixed parameters, the feedback that's collected can be used to inform future updates and training cycles

- **Corrections and teaching**:

  - **User corrections**: When users correct the model's outputs, this information can be valuable data for developers. It can show where the model is falling short and guide adjustments or provide direct learning signals in models designed to learn from interaction.

- **Active learning**: In some setups, when a user corrects a model's output, the model can use this correction as a learning instance, immediately adjusting its behavior for similar prompts in the future.

- **Personalization**:

  - **Adaptive responses**: Throughout an interaction session, some language models can adapt their responses based on the user's previous inputs, allowing for a more personalized interaction

  - **User preferences**: Understanding and adapting to user preferences can help the model provide more relevant and customized content

- **Interface and experience**:

  - **User interface (UI) design**: The design of the platform through which users interact with the model (such as a chatbot interface or a coding assistant) can affect how users phrase their prompts and respond to the model's outputs

  - **Usability**: A well-designed UI can make it easier for users to provide clear prompts and understand how to correct or provide feedback on the model's responses

- **Challenges in user interaction**:

  - **Misuse**: Users may intentionally try to trick or prompt the model to generate harmful or biased content, and thus robust safety and moderation mechanisms are required

  - **User errors**: Users may inadvertently provide prompts that are ambiguous or lead to unexpected results, highlighting the need for models to handle a wide range of inputs gracefully

- **Research and development**:

  - **User studies**: Ongoing research includes studying how users interact with language models to understand the best ways to design interfaces and feedback mechanisms

  - **Interface innovation**: Developers are continually innovating on how users can guide and interact with models, including using voice, gestures, or even brain-computer interfaces

- **The impact of user interaction**:

  - **Model improvement**: While the current version of GPT-4 doesn't learn from each interaction in real time, aggregated user interactions can inform developers and contribute to subsequent iterations of the model

  - **Customization and accessibility**: User interaction data can help make language models more accessible and useful to a broader audience, including individuals with disabilities or non-native speakers

User interaction is a dynamic and integral part of the language model ecosystem. The way users engage with models such as GPT-4 determines not only the immediate quality of the outputs but also shapes the future development of these AI systems. User feedback and interaction patterns are invaluable for refining the model's performance, enhancing user experience, and ensuring that the model serves the needs and expectations of its diverse user base.

In the next section, we'll cover RNNs in great detail. After, we'll compare the powerful Transformer model against RNNs.

Recurrent neural networks (RNNs) and their limitations

RNNs are a class of artificial neural networks that were designed to handle sequential data. They are particularly well-suited to tasks where the input data is temporally correlated or has a sequential nature, such as time series analysis, NLP, and speech recognition.

Overview of RNNs

Here are some essential aspects of how RNNs function:

- **Sequence processing**: Unlike feedforward neural networks, RNNs have loops in them, allowing information to persist. This is crucial for sequence processing, where the current output depends on both the current input and the previous inputs and outputs.

- **Hidden states**: RNNs maintain hidden states that capture temporal information. The hidden state is updated at each step of the input sequence, carrying forward information from previously seen elements in the sequence.

- **Parameters sharing**: RNNs share parameters across different parts of the model. This means that they apply the same weights at each time step, which is an efficient use of model capacity when dealing with sequences.

Limitations of RNNs

Despite their advantages for sequence modeling, RNNs have several known limitations:

- **Vanishing gradient problem**: As the length of the input sequence increases, RNNs become susceptible to the vanishing gradient problem, where gradients become too small for effective learning. This makes it difficult for RNNs to capture long-range dependencies in data.

- **Exploding gradient problem**: Conversely, gradients can also become too large, leading to the exploding gradient problem, where weights receive updates that are too large and the learning process becomes unstable.

- **Sequential computation**: The recurrent nature of RNNs necessitates sequential processing of the input data. This limits the parallelization capability and makes training less efficient compared to architectures such as **convolutional neural networks** (**CNNs**) or Transformers, which can process inputs in parallel.

- **Limited context**: Standard RNNs have a limited context window, making it difficult for them to remember information from the distant past of the sequence. This is particularly challenging in tasks such as language modeling, where context from much earlier in the text can be important. Also, there's limited memory capacity, which is a model's restricted ability to retain and process large amounts of information simultaneously.

Addressing the limitations

Several methods have been developed to address the limitations of RNNs:

- **Gradient clipping**: This technique is used to prevent the exploding gradient problem by capping the gradients during backpropagation to a maximum value.

- **Long short-term memory** (**LSTM**): LSTM is a type of RNN that's designed to remember information for long periods. It uses gates to control the flow of information and is much better at retaining long-range dependencies.

- **Gated recurrent unit** (**GRU**): GRUs are similar to LSTMs but with a simplified gating mechanism, which makes them easier to compute and often faster to train.

- **Attention mechanisms**: Although not a part of traditional RNNs, attention mechanisms can be used in conjunction with RNNs to help the model focus on relevant parts of the input sequence, which can improve performance on tasks that require an understanding of long-range dependencies.

While RNNs have been fundamental in the progress of sequence modeling, their limitations have led to the development of more advanced architectures such as LSTMs, GRUs, and the Transformer, which can handle longer sequences and offer improved parallelization. Nonetheless, RNNs and their variants remain a crucial topic of study and application in the field of deep learning.

Comparative analysis – Transformer versus RNN models

When comparing Transformer models to RNN models, we're contrasting two fundamentally different approaches to processing sequence data, each with its unique strengths and challenges. This section will provide a comparative analysis of these two types of models:

- **Performance on long sequences**: Transformers generally outperform RNNs on tasks involving long sequences because of their ability to attend to all parts of the sequence simultaneously

- **Training speed and efficiency**: Transformers can be trained more efficiently on hardware accelerators such as GPUs and TPUs due to their parallelizable architecture

- **Flexibility and adaptability**: Transformers have shown greater flexibility and have been successfully applied to a wider range of tasks beyond sequence processing, including image recognition and playing games

- **Data requirements**: RNNs can sometimes be more data-efficient, requiring less data to reach good performance on certain tasks, especially when the dataset is small

Let's consider the current landscape:

- **Dominance of transformers**: In many current applications, particularly in NLP, Transformers have largely supplanted RNNs due to their superior performance on a range of benchmarks.

- **The continued relevance of RNNs**: Despite this, RNNs and their more advanced variants, such as LSTMs and GRUs, continue to be used in specific applications where model size, computational resources, or data availability are limiting factors.

In conclusion, while both Transformers and RNNs have their place in the toolkit of ML models, the choice between them depends on the specific requirements of the task, the available data, and computational resources. Transformers have become the dominant model in many areas of NLP, but RNNs still maintain relevance for certain applications and remain an important area of study.

Summary

Language models such as GPT-4 are built on a foundation of complex neural network architectures and processes, each serving critical roles in understanding and generating text. These models start with extensive training data encompassing a diverse array of topics and writing styles, which is then processed through tokenization to convert text into a numerical format that neural networks can work with. GPT-4, specifically, employs the Transformer architecture, which eliminates the need for sequential data processing inherent to RNNs and leverages self-attention mechanisms to weigh the importance of different parts of the input data. Embeddings play a crucial role in this architecture by converting words or tokens into vectors that capture semantic meaning and incorporate the order of words through positional embeddings.

User interaction significantly influences the performance and output quality of models such as GPT-4. Through prompts, feedback, and corrections, users shape the context and direction of the model's outputs, making it a dynamic tool capable of adapting to various applications and tasks. Ethical considerations and the implementation of safety and moderation systems are also paramount, addressing issues such as bias, misinformation, and the potential impact on jobs. These concerns are mitigated through strategies such as content filtering, RLHF, and ongoing research to improve the model's robustness and trustworthiness. As the use of language models expands across industries and applications, these considerations ensure that they remain beneficial and ethical tools in advancing human-computer interaction.

In the next chapter, we'll build upon what we learned about LLM architecture in this chapter and explore how LLMs make decisions.

2

How LLMs Make Decisions

How LLMs make decisions is extremely complex, but it's something you should be aware of. In this chapter, we will provide you with a comprehensive examination of the decision-making processes in LLMs, starting with an analysis of how these models use probability and statistics to process information and predict outcomes. We will then explore the complex methodology LLMs employ to interpret inputs and construct responses. Furthermore, we will address the challenges and limitations that are inherent in LLMs, such as bias and reliability issues. We will also touch upon the current state and potential difficulties in ensuring the accuracy and fairness of these models. In the concluding part of this chapter, we will discuss the progressive methods and prospective advancements in the field of LLMs, signifying a dynamic area of technological development.

In this chapter, we'll cover the following main topics:

- Decision-making in LLMs – probability and statistical analysis
- From input to output – understanding LLM response generation
- Challenges and limitations in LLM decision-making
- Evolving decision-making – advanced techniques and future directions

By the end of this chapter, you will understand how the decision-making process is implemented in LLMs.

Decision-making in LLMs – probability and statistical analysis

Decision-making in LLMs involves complex algorithms that process and generate language based on a variety of factors. These include the input data they were trained on, the specific instructions or prompts they receive, and the statistical models that underlie their programming.

In this section, we'll provide an overview of how LLMs use probability and statistical analysis in decision-making.

Probabilistic modeling and statistical analysis

Probabilistic modeling is a cornerstone of how LLMs such as GPT-4 function. This approach allows the model to process natural language so that it reflects the complexities and variances inherent in human language use. Let's take a deeper look at several aspects of probabilistic modeling in LLMs:

- **Fundamentals of probabilistic modeling**: Probabilistic modeling is based on the concept of probability theory, which is used to model uncertainty. In the context of LLMs, this means that the model doesn't just learn fixed rules of language; instead, it learns the likelihood of certain words or phrases following others.

- **Sequence modeling with neural networks**: LLMs are a type of sequence model. They are designed to handle sequential data, such as text, where the order of the elements is crucial. For each potential next word in a sequence, the model generates a probability distribution while considering the words that have come before. This distribution reflects the model's "belief" about which words are most likely to come next. When generating text, the model samples from this distribution.

- **The Transformer architecture**: The Transformer, a type of neural network architecture, as discussed in the previous chapter, is particularly well-suited to this kind of probabilistic modeling because of its attention mechanisms. These mechanisms allow the model to weigh different parts of the input text when predicting the next word. It can "pay attention" to the entire context or focus on certain relevant parts, which is crucial for understanding the nuances of language.

- **Training on data and patterns**: During training, LLMs are fed huge amounts of text and learn to predict the probability of a word given the previous words in a sentence. This process, which was covered in the previous chapter, is not just about the frequency of word sequences but also about their context and usage patterns.

- **Softmax function**: A key component of the probabilistic model in LLMs is the softmax function. It takes the raw outputs of the model (which can be thought of as scores) and turns them into a probability distribution over the potential next words.

- **Loss function and optimization**: During training, a loss function measures how well the model's predictions match the actual outcomes. The model is optimized using algorithms such as stochastic gradient descent to minimize this loss, which involves adjusting the model's parameters to improve its probability estimates.

- **Handling ambiguity**: One of the challenges in probabilistic modeling for language is handling ambiguity. Words can have multiple meanings, and phrases can be interpreted in different ways, depending on the context. LLMs use the statistical patterns learned from data to handle this ambiguity, choosing the most probable meaning based on the context.

- **Model fine-tuning**: After its initial training, an LLM can be fine-tuned on more specific datasets. This allows the model to adjust its probabilistic predictions to better fit particular domains or styles of language.

- **Limitations and challenges**: While probabilistic modeling is powerful, it has its limitations. LLMs can sometimes generate text that is statistically probable but doesn't make sense or is factually incorrect. This is an area of active research as developers seek to improve the model's understanding and generation capabilities.

Probabilistic modeling in LLMs represents a significant advancement in the field of NLP, enabling these models to generate text that is often indistinguishable from that written by humans. The continuous refinement of these probabilistic methods is a key area of development that aims to achieve ever-more sophisticated levels of language understanding and generation.

Training on large datasets

As discussed previously, during training, LLMs are fed huge amounts of text and learn to predict the probability of a word given the previous words in a sentence. This process, which was covered in the previous chapter, is not just about the frequency of word sequences but also about their context and usage patterns.

Contextual understanding

Contextual understanding in LLMs such as GPT-4 is one of the most critical aspects of their operation. It allows them to interpret and respond to inputs in a way that is relevant and coherent. Let's take a closer look at how LLMs achieve this:

- **Understanding context through patterns**: As LLMs are trained on large amounts of text data, they learn patterns of language usage. This training enables them to pick up on the context in which words and phrases are typically used. For example, the word "apple" might be understood as a fruit in one context or as a technology company in another, depending on the surrounding words.

- **Attention mechanisms**: The Transformer architecture employs attention mechanisms to enhance contextual understanding. These mechanisms allow the model to focus on different parts of the input sequence, weighing them according to their relevance to the current task. This is how the model can consider the entire context of a sentence or paragraph when deciding which words to generate next.

- **Embeddings and positional encodings**: As discussed previously, LLMs use embeddings to convert words and tokens into numerical vectors that capture their meaning. These embeddings are context-dependent and can change based on the position of a word in a sentence, thanks to positional encodings. This is how the word "bank" can have different meanings when used in different contexts – for example, "river bank" and "money bank."

- **Layered understanding**: LLMs typically have multiple layers, with each layer capturing different aspects of language. Lower layers might focus on the syntax and grammar, while upper layers capture higher-level semantic meaning. This allows the model to process input at various levels of complexity, from basic word order to nuanced implications and inferences.

- **Handling ambiguity and polysemy**: Ambiguity is a natural part of language, and words can have multiple meanings (polysemy). LLMs use the context provided by the user to disambiguate words and phrases. For instance, if a user asks about "taking a break," the model understands this in the context of resting rather than "breaking something" due to the surrounding words that imply rest.

- **Calculating probabilities**: Statistical analysis in an LLM involves calculating probabilities for different potential outputs. The context is crucial for this process; for instance, if a user is discussing a topic such as climate change, the model uses the context to give higher probabilities to words and phrases related to that topic.

- **Continuous learning**: While LLMs are not capable of learning in real-time post-deployment in the same way humans do, some systems are designed to update their models periodically with new data, allowing them to adapt to changes in language use over time.

- **Limitations and challenges**: Despite these sophisticated mechanisms, LLMs still face challenges in contextual understanding. They can misunderstand nuances, fail to grasp sarcasm or idiomatic expressions, and generate nonsensical or off-topic responses if the context is too complex or too subtle.

- **Ethical considerations**: As mentioned previously, contextual understanding also brings ethical considerations. LLMs might inadvertently generate biased or sensitive content if the context cues are misinterpreted. It is an ongoing challenge to ensure that the models are as fair and unbiased as possible.

- **Applications**: In practical applications, contextual understanding is crucial. It enables LLMs to perform tasks such as translation, summarization, and question-answering with a high degree of accuracy and relevance.

The decision-making process in LLMs regarding contextual understanding is an active area of research and development, with each new model iteration bringing improvements that enable more sophisticated interactions with human users.

Machine learning algorithms

Machine learning (**ML**) algorithms form the backbone of LLMs, leveraging a variety of statistical techniques to process and generate language. Let's take a closer look at the most pertinent algorithms and methods that are used:

- **Supervised learning**: LLMs often use supervised learning, where the model is trained on a labeled dataset. For language models, the "labels" are typically the next few words in a sequence. The model learns to predict these labels (words) based on the input it receives.

- **Regression analysis**: In the context of LLMs, regression analysis isn't used in the traditional sense of fitting a line to data points. Instead, it's a broader class of algorithms that the model uses to map input features (words or tokens) to continuous output variables (the embeddings or the logits that will be turned into probabilities for the next word).

- **Bayesian inference**: Bayesian inference allows the model to update its predictions based on new data, incorporating the concept of probability to handle uncertainty. In LLMs, this method is not typically used in real time but can be a part of the training process, particularly in models that incorporate elements of unsupervised learning or reinforcement learning.

- **Gradient descent and backpropagation**: These are the most common algorithms that are used to train neural networks, including LLMs. Gradient descent searches for the minimum value of the loss function – a measure of how far the model's predictions are from the actual outcomes. Backpropagation is used to calculate the gradient of the loss function concerning each parameter in the model, allowing for efficient optimization.

- **Stochastic gradient descent** (**SGD**): A variant of gradient descent, SGD updates the model's parameters using only a small subset of the data at a time, which makes the training process much faster and more scalable for large datasets.

- **Transformer models**: The Transformer model, as covered previously, uses self-attention mechanisms to weigh the influence of different parts of the input data. This allows the model to focus more on certain parts of the input when making predictions.

- **Regularization techniques**: To prevent overfitting – the phenomenon of a model performing well on the training data but poorly on that data it has not seen –LLMs employ regularization techniques. These include methods such as dropout, where random subsets of neurons are "dropped out" during training to increase the robustness of the model.

- **Transfer learning**: Transfer learning involves taking a model that has been trained on one task and fine-tuning it on a different, but related, task. This is common practice with LLMs, where a model that's been pre-trained on a massive corpus of text is later fine-tuned for specific applications.

- **Reinforcement learning** (**RL**): Some LLMs integrate RL, where the model learns to make decisions by receiving rewards or penalties. This is less common in standard LLM training but can be used in specific scenarios, such as dialog systems, where user feedback is available.

- **Neural architecture search** (**NAS**): NAS is a process by which an ML algorithm searches for the best neural network architecture. This is an advanced technique that can be used to optimize LLMs for specific tasks or efficiency.

- **Data augmentation techniques**: These techniques involve creating additional training data from the existing data through various transformations, enhancing the model's ability to generalize and perform better on unseen data.

- **Attention techniques**: Various attention mechanisms, including self-attention and multi-head attention, allow the model to focus on different parts of the input data, enhancing its ability to understand and generate coherent and contextually relevant text.

- **Evaluation metrics**: Lastly, ML algorithms in LLMs rely on various evaluation metrics to measure their performance. These include perplexity, the BLEU score for translation tasks, the F1 score for classification tasks, and many others, depending on the specific application.

Collectively, these algorithms and techniques enable LLMs to process language at a high level, allowing them to generate text that is coherent, contextually relevant, and often indistinguishable from text written by humans. However, they also require careful tuning and a deep understanding of both the algorithms themselves and the language data they are trained on.

Feedback loops

Feedback loops in ML, including in the context of LLMs, are mechanisms by which the model's performance is assessed and improved over time through interaction with its environment or users. Let's take a closer look at how feedback loops operate within LLMs:

- **Types of feedback loops**:

  - **Supervised learning feedback loop**:

    - In a supervised learning setting, the feedback loop involves training the model on a dataset where the correct output is known (the "label"), and the model's predictions are compared to these labels

    - The model receives feedback in the form of loss gradients, which tell it how to adjust its parameters to make better predictions in the future

- **RL feedback loop:**

  - In RL, the feedback comes in the form of rewards or penalties, often referred to as positive or negative reinforcement.

  - An LLM might be used in an interactive setting where it generates responses to user inputs. If the response leads to a successful outcome (for example, user satisfaction), the model receives positive feedback; if not, it receives negative feedback.

- **Mechanisms of feedback:**

  - **Backpropagation:** In most neural network training, including LLMs, backpropagation is used to provide feedback. This is a method by which the model learns from errors by propagating them back through the network's layers, adjusting the weights accordingly.

  - **Reward functions:** In RL, a reward function provides feedback to the model based on the actions it takes. For instance, in a conversational AI setting, longer user engagement might result in higher rewards.

  - **User interaction:** As mentioned previously, user interaction can be a source of feedback, especially for models deployed in the real world. User corrections, time spent on a generated article, click-through rates, and other metrics can serve as feedback.

- **Continuous improvement:**

  - **Model retraining:** Models can be retrained with new data that includes past mistakes and successes, allowing them to update their parameters and improve over time

  - **Fine-tuning:** Models may also be fine-tuned on specific tasks or datasets based on feedback, which is a more targeted approach than full retraining

  - **Active learning:** Some systems use active learning, where the model identifies areas where it is uncertain and requests feedback in the form of new data or human input to improve

- **Challenges and considerations:**

  - **Feedback quality:** The quality of feedback is crucial. Poor feedback can lead to incorrect learning and reinforce biases or undesirable behaviors.

  - **Feedback loop dynamics:** Feedback loops can become problematic if they start to reinforce themselves in negative ways, such as amplifying biases or leading to echo chambers.

  - **Ethical and safety concerns:** Ensuring that feedback doesn't lead to the development of unsafe or unethical behaviors in LLMs is an ongoing challenge in AI safety and ethics.

Feedback loops are essential for the adaptive and predictive capabilities of LLMs, allowing them to refine their decision-making and language understanding continually. They are particularly important in applications where LLMs interact with users in dynamic environments, such as chatbots, personal assistants, or interactive storytelling.

Uncertainty and error

Uncertainty and error are intrinsic to any statistical model, including LLMs such as GPT-4. In this section, we'll take an in-depth look at how LLMs deal with these issues.

The nature of uncertainty in LLMs

In understanding the intricacies of LLMs, three fundamental concepts are pivotal:

- **Probabilistic nature**: The core of LLMs is probabilistic; they generate language based on a distribution of possible next words or tokens. This means that the model's output is inherently uncertain, and the model must estimate many possible outcomes.

- **Context sensitivity**: LLMs rely heavily on context to make predictions. If the context is unclear or ambiguous, the model's uncertainty increases, which can lead to errors in the output.

- **Data sparsity**: No matter how large the training dataset is, there will always be gaps. When LLMs encounter scenarios that were underrepresented or not present in their training data, they may be less certain about the correct output.

How LLMs handle uncertainty

To grasp how LLMs generate and refine their outputs, it's essential to consider various key mechanisms:

- **Softmax function**: When generating text, the model uses a softmax function to convert the logits (the raw output from the last layer of the neural network) into a probability distribution. The word with the highest probability is typically selected as the next word in the sequence.

- **Sampling strategies**: Instead of always choosing the most likely next word, LLMs can use different sampling strategies to introduce variety into the text they generate or to explore less likely, but potentially more interesting, paths.

- **Beam search**: In tasks such as translation, LLMs might use a beam search algorithm to consider multiple potential translations at once and select the most probable overall sequence, rather than making decisions word by word.

- **Uncertainty quantification**: Some models are capable of quantifying their uncertainty, which can be useful for flagging when the model's output should be treated with caution.

- **Monte Carlo dropout**: This technique is used during inference to provide a measure of uncertainty in the model's predictions. It does this by randomly dropping out different parts of the network and sampling multiple times, which helps in understanding the variability and reliability of the model's output.

Error types and sources

Addressing the accuracy and reliability of LLMs involves understanding the following nuances:

- **Systematic errors**: These occur when the model consistently misinterprets certain inputs due to biases or flaws in the training data.

- **Random errors**: These occur unpredictably and are usually due to the inherent randomness in the model's probability estimates.

- **Overfitting and underfitting**: Overfitting occurs when a model is too closely tailored to the training data and fails to generalize to new data. Underfitting occurs when the model is too simple to capture the complexity of the training data.

- **Model misinterpretation**: Errors can arise when users misinterpret the capabilities of the model, expecting it to have an understanding or abilities beyond its actual capacity.

Error mitigation strategies

In the pursuit of optimizing LLMs, techniques such as the ones mentioned here play crucial roles in enhancing performance and maintaining relevance over time:

- **Regularization**: Techniques such as dropout are used during training to prevent overfitting and help the model generalize better to new data

- **Ensemble methods**: Using a collection of models to make a decision can reduce the impact of errors as the models can correct each other's mistakes

- **Human-in-the-loop**: For critical applications, human oversight can be used to review and correct the model's output

- **Continuous training**: Continually updating the model with new data can help it learn from past errors and adapt to changes in language use over time

Ethical and practical implications

The following aspects are fundamental in managing the deployment and user interaction process regarding LLMs:

- **Trust**: Users need to understand the probabilistic nature of LLMs to set appropriate expectations for their reliability

- **Safety**: In high-stakes scenarios, the potential for error must be managed carefully to avoid harmful outcomes

- **Transparency**: Users must be aware of how LLMs make decisions and the potential for uncertainty and error in their outputs

In summary, while LLMs have advanced considerably, they are not infallible and their outputs must be evaluated critically, especially when used in sensitive or impactful contexts. Understanding the nature of uncertainty and error in these models is crucial for both users and developers to use them effectively and ethically.

From input to output – understanding LLM response generation

The process of generating a response in an LLM such as GPT-4 is a complex journey from input to output. In this section, we'll take a closer look at the steps that are involved.

Input processing

The following are the key preprocessing steps in LLMs:

1. **Tokenization**: Splitting the text into tokens based on predefined rules or learned patterns.
2. **Embedding**: Sometimes, tokens are normalized to a standard form. For instance, "USA" and "U.S.A." might be normalized to a single form.
3. **Positional encoding**: Each unique token is associated with an index in a vocabulary list. The model will use these indices, not the text itself, to process the language.

Model architecture

The following are central components in the architecture of LLMs:

- **Transformer blocks**: Each Transformer block contains two main parts: a multi-head self-attention mechanism and a position-wise feed-forward network.

- **Self-attention**: As mentioned previously, the attention mechanism allows the model to weigh the importance of different tokens when predicting the next word. It can focus on the entire input sequence and determine which parts are most relevant at any given time.

Decoding and generation

The process of decoding and generation in the context of LLMs such as GPT-4 involves several intricate steps that convert a given input into a coherent and contextually appropriate output. This process is the core of how these models communicate and generate text. Let's take a closer look at each step.

The probability distribution process involves the following aspects:

- **Logits**: Splitting the text into tokens based on predefined rules or learned patterns.

- **Softmax layer**: Sometimes, tokens are normalized to a standard form.

- **Temperature**: Each unique token is associated with an index in a vocabulary list. The model will use these indices, not the text itself, to process the language.

Output selection is comprised of the following components:

- **Greedy decoding**: The most straightforward selection method is greedy decoding, where the model always picks the word with the highest probability as the next token. This approach is deterministic.

- **Beam search**: Beam search is a more nuanced technique where the model keeps track of multiple sequences (the "beam width") and extends them one token at a time, ultimately choosing the sequence with the highest overall probability.

- **Random sampling**: The model can also randomly sample from the probability distribution, which introduces randomness into the output and can lead to more creative and less predictable text.

- **Top-k sampling**: This method restricts the sampling pool to the k most likely next words. The model then samples only from this subset, which can lead to a balance between variety and coherence.

- **Top-p (nucleus) sampling**: Instead of picking a fixed number of words, top-p sampling chooses from the smallest set of words whose cumulative probability exceeds a threshold, p. This focuses on a "nucleus" of likely words, ignoring the long tail of the distribution.

The challenges in decoding and generation

Let's take a closer look at the challenges we must overcome:

- **Repetitiveness**: Even sophisticated models can fall into repetitive loops, especially with greedy decoding methods

- **Coherence over long texts**: Maintaining coherence over longer texts is challenging as the model must remember and appropriately reference information that may have been introduced much earlier

- **Context limitations**: There is a limit to how much context the model can consider, known as the context window, which can affect the quality of the generated text for inputs that exceed this window

Future directions

Now, let's consider some future directions:

- **Attention span**: Research is ongoing into models that can handle longer contexts, either through modifications to the attention mechanism or different approaches to memory

- **Adaptive decoding**: Adapting the decoding strategy based on the type of text being generated (for example, creative writing versus technical instructions) could improve the quality of the generated text

- **Feedback-informed generation**: Incorporating real-time feedback loops could help models adjust their generation process on the fly, leading to more interactive and adaptive communication

Decoding and generation is a field of active research, with each new model version aiming to produce more accurate, coherent, and contextually rich outputs. This not only involves improvements to the underlying algorithms but also a better understanding of how humans use language.

Iterative generation

Iterative generation is a fundamental process that's used by LLMs such as GPT-4 to produce text. This process is characterized by two main components: the autoregressive process and the establishment of a stop condition. Iterative generation is a multi-step process that may involve revisions, while decoding and generation are generally one-pass processes. Let's take a closer look.

Autoregressive process

Over time, the following critical aspects dictate how LLMs process and generate language:

- **Sequential predictions**: In an autoregressive model, each output token (which could be a word or part of a word) is predicted sequentially. The prediction of each subsequent token is conditional on the tokens that have been generated so far.

- **Dependency on previous tokens**: The model's prediction at each step is based on all the previous tokens in the sequence, which means that the model "remembers" what it has already generated. This is crucial for maintaining coherence and context.

- **Latent representations**: As tokens are generated, the model updates its representations of the sequence's meaning internally. These representations are complex vectors in high-dimensional space that encode the semantic and syntactic nuances of the text.

- **Complexity over time**: With each new token, the complexity of the text increases. The model must balance various factors, such as grammar, context, style, and the specific requirements of the task at hand.

Stop condition

These are mechanisms in LLMs that guide when and how to conclude the generation of text:

- **End-of-sequence token**: Many LLMs use a special token to signify the end of a sequence, often referred to as `<EOS>` or `[end]`. When the model predicts this token, the iterative generation process stops.

- **Maximum length**: To prevent runaway generation, a maximum sequence length is often set. Once the generated text reaches this length, the model will stop generating new tokens, regardless of whether it has reached a natural conclusion.

- **Task-specific conditions**: For certain applications, there might be other conditions that determine when the generation process should stop. For example, in a question-answering task, the model might be programmed to stop after generating a sentence that appears to answer the question.

Challenges in iterative generation

Here are some challenges you should consider:

- **Repetition**: Models may get stuck in loops, repeating the same phrase or structure. This can often be mitigated by modifying the sampling strategy or by using techniques such as deduplication post-generation.

- **Context dilution**: As more tokens are generated, the influence of the initial context can diminish, potentially leading to a loss of coherence.

- **Computational efficiency**: Generating text token by token can be computationally intensive, particularly for longer sequences or when using sampling strategies that require many potential continuations to be evaluated.

Future directions

Advancements in the design of LLMs aim to improve the following areas:

- **Longer context windows**: Researchers are working on expanding the context window that LLMs can consider, allowing for better maintenance of context over longer texts

- **Efficient decoding**: Newer models and techniques are being developed to generate text more efficiently, balancing the trade-offs between speed, coherence, and diversity

- **Interactive generation**: Some research focuses on making the generation process interactive, allowing users to guide the generation in real time or provide feedback that the model can incorporate immediately

Iterative generation is at the core of how LLMs such as GPT-4 produce text, enabling them to create everything from simple sentences to complex narratives and technical documents. Despite its challenges, the autoregressive nature of LLMs is what allows text to be generated that is often indistinguishable from that written by humans. As research progresses, we can expect to see more sophisticated models that handle the complexities of language with even greater finesse.

Post-processing

Post-processing is a crucial step in the workflow of text generation with LLMs, which ensures that the raw output from the model is polished and made presentable for the intended audience or application. Let's take a detailed look at the components of post-processing.

Detokenization

After an LLM generates a sequence of tokens, they must be converted back into a format that can be understood and read by humans. This process is known as detokenization. Let's take a look at what's involved:

- **Joining tokens**: Tokens that represent subparts of words or punctuation need to be joined together correctly. For example, "New," "##York," and "City" would need to be detokenized to "New York City."

- **Whitespace management**: Adding spaces between words is generally straightforward but can be complex with languages that don't use whitespace in the same way as English or when dealing with special characters and punctuation.

- **Special tokens**: The model might generate special tokens that indicate formatting or other non-standard text elements. These need to be interpreted or removed during detokenization.

Formatting

Once the text has been detokenized, it may need additional formatting to ensure it meets the required standards for grammar, style, and coherence. This can involve several processes:

- **Grammar checks**: Automated grammar checkers can identify and correct basic grammatical errors that the LLM may have produced.

- **Style guides**: For certain applications, the text might need to adhere to specific style guides. This could involve adjusting word choice, sentence structure, or punctuation.

- **Custom rules**: Some applications may require specific formatting rules, such as capitalizing certain words, formatting dates and numbers, or adding hyperlinks.

- **Domain-specific adjustments**: Technical, legal, or medical texts might require additional checks to ensure terminology and formatting meet industry standards.

Challenges in post-processing

In managing the output quality of LLMs, the following issues are critical to address:

- **Loss of meaning**: Incorrect detokenization can sometimes change the meaning of the text or render it nonsensical

- **Overcorrection**: Automated grammar and style correction tools might "overcorrect" the text, making changes that don't align with the intended meaning or style

- **Scalability**: Post-processing needs to be efficient to handle large volumes of text without introducing significant delays

Future directions

The following are essential strategies for elevating the quality and effectiveness of text generated by LLMs:

- **ML in post-processing**: ML models specifically trained for post-processing tasks can improve the quality of the output text

- **User feedback integration**: Incorporating user feedback into post-processing can help tailor the text to the preferences of the audience

- **Adaptive formatting**: Developing systems that can adapt the formatting based on the context and intended use of the text can enhance the readability and impact of the generated content

Post-processing is the final touch that transforms the model's output into polished, user-friendly content. It is an area where even small improvements can significantly enhance the usability of LLM-generated text, making it more accessible and effective for the task at hand.

Challenges and limitations in LLM decision-making

LLMs such as GPT-4 are technological marvels, but they come with a set of challenges and limitations that impact their decision-making abilities. Here are some of the challenges and limitations we must consider:

- **Understanding context and nuance**:

  - **Ambiguity**: LLMs may struggle with ambiguity in language. They sometimes cannot determine the correct meaning of a word or phrase without clear context.

  - **Sarcasm and irony**: Detecting sarcasm or irony is particularly challenging because it often requires understanding subtle cues and having a deep cultural context that LLMs may not have.

  - **Long-term context**: Maintaining coherence over long conversations or documents is difficult as LLMs might lose track of earlier context.

- **Generalization versus specialization**:

  - **Overfitting**: LLMs can become too specialized to the training data, making them less able to generalize to new types of data or problems

  - **Underfitting**: Conversely, LLMs might not capture the specifics of certain tasks or domains if they generalize too much

- **Data bias and fairness**:

  - **Training data bias**: LLMs reflect the biases in their training data, which can lead to unfair or prejudiced outcomes

  - **Representation**: If the training data doesn't represent the diversity of language and communication styles, the LLM's performance can be uneven across different user groups

- **Ethical and moral reasoning**:

  - **Value alignment**: LLMs don't possess human values and can generate ethically questionable content

  - **Moral decision-making**: LLMs cannot make moral decisions or understand ethical nuances in the way humans do

- **Reliability and error rates**:

  - **Inconsistencies**: LLMs might produce inconsistent or contradictory information, especially when generating information over multiple sessions

  - **Factuality**: LLMs can confidently present incorrect information as fact, leading to misinformation if it's not checked

- **Interpretability and transparency**:

  - **Black box nature**: An LLM's decision-making process is complex and often not easily interpretable, which can make it hard to understand why it generates certain outputs

  - **Transparency**: It can be difficult to provide clear explanations for the model's behavior, which is a significant issue for accountability

- **Computational and environmental costs**:

  - **Resource intensive**: Training and running LLMs requires a considerable amount of computational resources, which leads to high energy consumption and environmental impact

  - **Scalability**: The computational cost also affects scalability as deploying LLMs to many users can be resource-prohibitive

- **Dependence on human oversight**:

  - **Supervision needs**: Many LLM applications require human oversight to ensure the quality and appropriateness of outputs

  - **Feedback loop limitations**: While feedback loops can improve LLMs, they can also perpetuate errors if they're not managed carefully

- **Safety and security**:

  - **Robustness**: LLMs can be sensitive to adversarial attacks where small, carefully crafted changes to the input can lead to incorrect outputs

  - **Manipulation**: There's a risk of LLMs being used to generate manipulative content, such as deepfakes or spam

- **Societal impact**:

  - **Job displacement**: Automating tasks that LLMs can perform may lead to the displacement of jobs, raising societal and economic concerns

  - **Digital divide**: The benefits of LLMs may not be evenly distributed, potentially exacerbating the digital divide

Despite these challenges and limitations, LLMs represent a significant step forward in AI and natural language processing. Continuous research is directed toward mitigating these issues, improving the models' decision-making processes, and finding ways to use LLMs responsibly and effectively. It's a dynamic field that requires not only technical innovation but also ethical and societal considerations.

Evolving decision-making – advanced techniques and future directions

The field of AI, particularly the branch that deals with LLMs, is rapidly evolving. The decision-making capabilities of these models are constantly being enhanced through advanced techniques and research into future directions. Let's explore some of these advancements and the potential paths that future developments might take.

Advanced techniques in LLM decision-making

Advancements in these domains are driving the evolution of LLMs, each contributing to more nuanced text processing and enhanced model performance:

- **Transformer architecture**: The Transformer architecture has been pivotal in the recent successes of LLMs. Innovations continue to emerge in how these models handle long-range dependencies and contextual information.

- **Sparse attention mechanisms**: To handle longer texts efficiently, researchers are developing sparse attention patterns that allow LLMs to focus on the most relevant parts of the input without being overwhelmed by data.

- **Capsule networks**: These are designed to enhance the model's ability to understand hierarchical relationships in data, potentially improving the decision-making process by capturing more nuanced patterns.

- **Energy-based models**: By modeling decision-making as an energy minimization problem, these models can generate more coherent and contextually appropriate responses.

- **Adversarial training**: This involves training models to resist adversarial attacks, which can improve their robustness and reliability.

- **Neuro-symbolic AI**: Combining deep learning with symbolic reasoning, neuro-symbolic AI could lead to models that have a better grasp of logic, causality, and common-sense reasoning.

Future directions for LLM decision-making

The future of LLMs is poised to be shaped by the following advancements:

- **Improved contextual understanding**: Future LLMs may incorporate mechanisms that allow for a more profound understanding of context, not just within a single conversation or document but across multiple interactions.

- **Continual learning**: Enabling LLMs to learn from new data continuously without forgetting previous knowledge is a significant goal. Techniques such as elastic weight consolidation are being explored to achieve this.

- **Interpretable AI**: There is a push toward making AI decision-making more interpretable and transparent. This includes developing models that can explain their reasoning and choices in human-understandable terms.

- **Enhanced common sense and world knowledge**: Future models might integrate structured world knowledge and common-sense reasoning databases, improving their decision-making capabilities significantly.

- **Biologically inspired AI**: Drawing inspiration from neuroscience, future LLMs might mimic the human brain's decision-making processes more closely, potentially leading to more natural and intuitive AI behavior.

- **Hybrid models**: Combining LLMs with other types of AI, such as reinforcement learning agents, could lead to systems that can both generate natural language and interact with the environment in sophisticated ways.

- **Ethical AI**: As LLMs become more advanced, ensuring they make decisions that align with human values and ethics becomes increasingly important. Research into ethical AI focuses on embedding moral decision-making processes within the model's architecture.

- **Personalization**: Personalizing responses based on user preferences and history, while maintaining privacy and security, is an area of active research.

- **Multimodal AI**: Integrating LLMs with other types of data, such as visual or auditory information, could lead to richer decision-making capabilities and more versatile applications.

- **Quantum computing**: Quantum algorithms have the potential to revolutionize LLMs by enabling them to process information in fundamentally new ways, though this is still in the exploratory stage.

- **Multilingual and cross-lingual capabilities**: Future LLMs are expected to enhance their ability to understand and generate text across multiple languages and leverage cross-lingual information, improving global accessibility and usability.

- **Sustainability and efficiency**: There is a growing focus on making LLMs more energy-efficient and environmentally sustainable by optimizing algorithms, reducing computational requirements, and exploring greener AI technologies.

Challenges and considerations

As LLMs and their decision-making processes evolve, there will be challenges, including computational demands, potential biases in AI behavior, privacy concerns, and the need for regulatory frameworks. There will also be a continuous need for multidisciplinary collaboration among computer scientists, ethicists, sociologists, and policymakers to guide the development of these advanced AI systems.

The evolution of LLM decision-making is an exciting and active area of AI research, with many promising directions and techniques under exploration. The future of LLMs is likely to see models that are not only more powerful in terms of raw computational ability but also more nuanced, ethical, and aligned with human needs and values.

Summary

In this chapter, we focused on the decision-making process of LLMs, which utilize a complex interplay of probabilistic modeling and statistical analysis to interpret and generate language. LLMs, such as GPT-4, are trained on extensive datasets, allowing them to predict the likelihood of word sequences within a given context. The Transformer architecture plays a crucial role in this process, with its attention mechanisms assessing different input text elements to produce relevant output. We further explored the nuances of LLM training, emphasizing the importance of context and patterns learned from data to refine the models' predictive capabilities.

By addressing the challenges LLMs face, we provided insight into issues such as bias, ambiguity, and the balancing act between overfitting and underfitting. We also touched on the ethical implications of AI-generated content and the continuous need for model fine-tuning to achieve more sophisticated language understanding. Looking ahead, we anticipate advancements in LLM decision-making, highlighting ongoing research in areas such as improved contextual understanding, continuous learning, and the integration of multimodal data. The evolution of LLMs is portrayed as a dynamic and collaborative field requiring both technical innovation and a strong consideration of ethical and societal impacts. At this point, you should have a comprehensive understanding of how the decision-making process is implemented in LLMs.

In the next chapter, we'll guide you through the mechanics of training LLMs, giving you a thorough grounding in creating effective LLMs.

Part 2:
Mastering LLM Development

In this part, you will learn about data, how to set up your training environment, hyperparameter tuning, and challenges in training LLMs. You will also learn about advanced training strategies, which entail transfer learning and fine-tuning, as well as curriculum learning, multitasking, and continual learning models. Instruction on fine-tuning LLMs for specific applications is also included; here, you will learn about the needs of NLP applications, tailoring LLMs for chatbots and conversational agents, customizing models for language translation, and fine-tuning for nuanced understanding. Finally, we will focus on testing and evaluation, which includes learning about metrics for measuring LLM performance, how to set up rigorous testing protocols, human-in-the-loop instances, ethical considerations, and bias mitigation.

This part contains the following chapters:

- *Chapter 3, The Mechanics of Training LLMs*
- *Chapter 4, Advanced Training Strategies*
- *Chapter 5, Fine-Tuning LLMs for Specific Applications*
- *Chapter 6, Testing and Evaluating LLMs*

3

The Mechanics of Training LLMs

Here, we will guide you through the intricate process of training LLMs, starting with the crucial task of data preparation and management. This process is fundamental to getting LLMs to perform in a desired way. We will further explore the establishment of a robust training environment, delving into the science of hyperparameter tuning and elaborating on how to address overfitting, underfitting, and other common training challenges, giving you a thorough grounding in creating effective LLMs.

In this chapter, we're going to cover the following main topics:

- Data – preparing the fuel for LLMs
- Setting up your training environment
- Hyperparameter tuning – finding the sweet spot
- Challenges in training LLMs – overfitting, underfitting, and more

By the end of this chapter, you should understand the roadmap for training LLMs, emphasizing the pivotal role of comprehensive data preparation and management.

Data – preparing the fuel for LLMs

Preparing datasets for the effective training of LLMs is a multi-step process that requires careful planning and execution. Here is a comprehensive guide on how to prepare datasets.

Data collection

Data collection is a fundamental step in the development of LLMs and involves gathering a vast and varied set of text data that the model will use to learn. The quality and diversity of this corpus are critical as they directly influence the model's ability to understand and generate language across different domains and styles. Let's take a look at an expanded view of the data collection process:

- **Scope of corpus**: The corpus should cover a wide range of topics to prevent the model from developing a narrow understanding of language. It should include literature from various genres, informative articles from different fields, dialogues from conversational datasets, technical documents, and other relevant text sources.

- **Language representation**: For multilingual models, the dataset must include texts in all target languages. It's important to ensure that less-resourced languages are adequately represented to avoid bias toward the more dominant languages.

- **Temporal diversity**: Including texts from different time periods can help the model understand language evolution and historical contexts, making it better at handling archaic terms and newer slang.

- **Cultural and demographic diversity**: The corpus should represent various cultural and demographic backgrounds to ensure that the model can understand and generate text that is inclusive and respectful of diversity.

- **Ethical compliance**: Data should be sourced from ethical channels, ensuring respect for copyright laws and intellectual property rights. This involves using texts that are in the public domain or obtaining appropriate licenses for protected content.

- **Legal compliance**: Comply with data privacy laws, such as GDPR or CCPA, especially when using texts that contain personal information. It's essential to anonymize and aggregate data where necessary to protect individual privacy.

- **Quality control**: Evaluate the quality of the texts to ensure they are free from errors and remove low-quality or spam content that could negatively influence the model's learning process.

- **Balanced representation**: Avoid overrepresentation of certain topics that could lead to biased predictions. Ensure that the model is exposed to a balanced view of sensitive subjects.

- **Data format and annotation**: Depending on the intended use of the LLM, the data may need to be annotated with additional information, such as part-of-speech tags or named-entity labels. The format should be consistent to facilitate efficient processing during training.

- **Data usage rights**: Secure the rights to use the data for **machine learning** (**ML**) purposes. This can involve negotiations and agreements with data providers, particularly for proprietary or commercial datasets.

- **Ongoing collection**: Data collection is not a one-time process; it's an ongoing activity that keeps the dataset up to date as languages evolve and new types of text emerge.

- **Source documentation**: Keep detailed records of where, when, and how data was collected. This documentation can be crucial for troubleshooting, audits, and reproducibility of research.

By meticulously collecting and curating the data, developers can create LLMs that are well rounded, less biased, and more reliable in their understanding and generation of language.

Data cleaning

Data cleaning is a critical phase in preparing datasets for training LLMs, as it directly impacts the model's ability to learn effectively. A more detailed look into the data cleaning process is as follows:

- **Correcting encoding issues**: Text data often comes from various sources, each potentially using different character encodings. It's essential to standardize the text to a consistent encoding format, such as UTF-8, to avoid character corruption. Tools such as `iconv` or programming libraries in Python can automate this process.

- **Removing noise**: Textual noise includes any irrelevant information that might confuse the model. This can be extraneous HTML tags, JavaScript code in web-scraped data, or corrupted text. Regular expressions and HTML parsers, such as Beautiful Soup, can help automate the removal of such noise.

- **Standardizing language**: Datasets may contain slang, abbreviations, or creative spellings. Depending on the model's intended use, you might want to standardize these to their full forms to ensure consistency.

- **Handling non-standard language**: If the dataset includes non-standard language elements, such as code snippets, mathematical formulas, or chemical equations, these should either be removed or systematically tagged if they are relevant to the model's tasks.

- **Anonymization**: **Personally identifiable information** (**PII**) must be detected and removed or anonymized to comply with privacy regulations. Techniques such as **named-entity recognition** (**NER**) can be used to identify PII, and various anonymization techniques can mask or remove this information.

- **Dealing with missing values**: In structured datasets, missing values can be problematic. Depending on the situation, you might fill them with placeholder values, interpolate them based on nearby data, or omit the entries altogether.

- **Unifying formats**: Dates, numbers, and other structured data should be converted to a uniform format. This can involve converting all dates to a standard format, such as YYYY-MM-DD, or ensuring all numbers are represented consistently.

- **Language correction**: Spelling errors and grammatical mistakes can be corrected using automated tools, such as spell checkers or language-parsing algorithms, although it's important to be cautious not to over-standardize and remove nuances important for certain tasks.

- **Duplicate removal**: Identifying and removing duplicate entries is important to prevent the model from giving undue weight to repeated information.

- **Data validation**: After cleaning, validate the dataset to ensure that the cleaning steps have been properly applied and that the data is in the correct format for model training.

- **Quality assessment**: Perform a quality assessment, possibly with human review, to ensure the data meets the standards required for effective LLM training.

- **Irrelevant or outdated information**: Removing or updating irrelevant or outdated information ensures the model is trained on accurate and current data, which enhances its relevance and performance.

Effective data cleaning not only improves the model's performance but also contributes to the fairness and ethical use of LLMs by preventing the learning of biases and ensuring the privacy of individuals represented in the data.

Tokenization

Tokenization is a pivotal preprocessing step in preparing data for training LLMs. It involves breaking down the text into smaller units, known as **tokens**, which can be words, subwords, or even individual characters. The choice of tokenization granularity has a significant impact on the model's subsequent training and performance.

Here are the major tokenization approaches:

- **Word-level tokenization**: This approach splits the text into words. It's straightforward and works well for languages with clear word boundaries, such as English. However, it can lead to a very large vocabulary size, which in turn may increase the model's complexity and resource requirements.

- **Subword tokenization**: Subword tokenization techniques, such as **byte-pair encoding** (**BPE**) or WordPiece, split words into smaller, more frequent pieces. This method can effectively reduce vocabulary size and handle out-of-vocabulary words by breaking them down into subword units. It strikes a balance between the flexibility of character-level models and the efficiency of word-level models. Subword tokenization is particularly useful for agglutinative languages where many morphemes combine to form a single word, or in cases where the model needs to handle a mix of different languages with varying morphologies.

- **Character-level tokenization**: In character-level tokenization, each character is treated as a separate token. This method ensures a small, fixed vocabulary size and allows the model to learn all the nuances of word formation. However, it can make learning long-range dependencies more challenging due to the increased sequence lengths.

- **Tokenization for specialized tasks**: For certain tasks, such as NER or part-of-speech tagging, tokenization might need to align with the linguistic properties of the text. Tokens may need to correspond to meaningful linguistic units, such as phrases or syntactic chunks.

- **Advanced techniques**: More recent approaches, such as SentencePiece or Unigram language model tokenization, don't rely on white space to determine token boundaries and can work well across multiple languages, including those without clear white space delimiters.

These are the considerations to take into account with tokenization:

- **Consistency**: It's important to apply the same tokenization method consistently across the entire dataset to prevent discrepancies that could hinder the model's learning process.

- **Handling special tokens**: LLMs often require special tokens to signify the start and end of sequences or to separate segments within the input. The tokenization process should incorporate these special tokens appropriately.

- **Alignment with downstream tasks**: The tokenization granularity should consider the end use of the LLM. For fine-grained tasks, such as translation or text generation, subword- or word-level tokenization might be preferable, while for character-level modeling of syntax or phonetics, character-level tokenization could be more appropriate.

Ultimately, the choice of tokenization impacts the model's ability to understand and generate language and should be carefully considered in the context of the specific goals and constraints of the LLM training project.

Annotation

Annotation, in the context of training LLMs for supervised learning tasks, is a meticulous process where the raw data is enriched with additional information that defines the correct output for a given input. This process allows the model to not only ingest the raw data but also to learn from the correct interpretations or classifications provided by these annotations. Let's get a deeper insight into this process:

- **Next-word prediction**: For tasks such as language modeling, data is annotated in a way that the model can learn to predict the next word in a sequence. This often involves shifting the sequence of tokens so that for each input token, the output token is the next word in the original text. The model learns to associate sequences of tokens with their subsequent tokens.

- **Sentiment analysis**: When preparing data for sentiment analysis, human annotators review text segments, such as sentences or paragraphs, and label them with sentiment scores or categories, such as positive, negative, or neutral. The precision of this annotation process is critical as it directly impacts the model's ability to correctly identify sentiment in new texts.

- **NER**: In NER tasks, annotators label words or phrases in the text that correspond to entities such as person names, organizations, locations, and so on. This labeling is often done using a tagging schema such as **beginning, inside, outside (BIO)**, which marks not just the entity, but also the position of the word within the entity.

- **Accuracy and consistency**: To ensure the model learns correctly, annotations must be accurate and consistent. This often involves creating a detailed annotation guideline that annotators can follow to reduce subjectivity and variance in the labeling process.

- **Annotation tools**: Specialized software tools are used to facilitate the annotation process. These tools can provide a user-friendly interface for annotators, automate parts of the annotation process with pre-annotations using heuristics or semi-supervised methods, and manage the workflow of large-scale annotation projects.

- **Quality control**: Implementing quality control mechanisms is essential. This may involve multiple annotators labeling the same data and using inter-annotator agreement metrics to ensure quality or having expert reviewers validate the annotations.

- **Handling ambiguity**: For ambiguous cases, it's important to either design the annotation guidelines to capture the ambiguity or have a strategy for resolving it, such as consensus among multiple annotators or deferring to expert judgment.

- **Scalability**: For LLMs, the annotation process must be scalable due to the large amounts of data required. This may involve crowdsourcing platforms or collaboration with professional data annotation companies.

- **Privacy considerations**: If the data being annotated contains personal or sensitive information, privacy-preserving measures must be taken, including data anonymization and securing the consent of the data subjects, if necessary.

Annotations are foundational for supervised learning as they provide the ground truth that the model strives to predict correctly. The quality of the training data annotations directly correlates with the performance of the LLM on the task it's being trained for.

Data augmentation

Data augmentation is an important technique in preparing datasets for training LLMs as it helps to create a more robust and generalizable model by artificially expanding the diversity of the training data. The following is a more in-depth explanation of some common data augmentation techniques:

- **Synthetic data generation**: This involves creating new data points from existing ones through various transformations. For text, this could mean using techniques such as random insertion, deletion, or swapping of words within a sentence while preserving grammatical correctness and meaning. Synonym replacement is another common method, where certain words are replaced with their synonyms.

- **Back translation**: This is a popular method for augmenting text data, especially in the context of machine translation. Here, a sentence is translated from one language to another (usually with an LLM) and then translated back to the original language. The round-trip translation process introduces linguistic variations, providing a form of paraphrasing that can help the model generalize better.

- **Noise injection**: Introducing noise into the data can make models more robust to variations and potential input errors. For textual data, this might involve adding typographical errors, playing with different casing, or inserting additional white space.

- **Paraphrasing**: Generating paraphrases of sentences or phrases can expand the dataset with diverse linguistic structures conveying the same meaning. Paraphrasing can be done using rule-based approaches or by employing models trained specifically for this task.

- **Data warping**: In the context of sequential data, such as text, warping can mean altering the sequence length by summarizing or expanding passages of text.

- **Using external datasets**: Incorporating data from external sources that are not part of the original dataset can also help in improving the diversity and size of the training corpus.

- **Translation augmentation**: For multilingual models, sentences can be translated into various languages and added to the dataset, increasing the model's exposure to different linguistic patterns.

- **Generative models**: Advanced data augmentation may utilize other generative models to create new data instances. For instance, **generative adversarial networks** (**GANs**) can be trained to generate text that is similar to human-written text.

- **Relevance to task**: The augmentation strategies chosen must be relevant to the task the LLM will perform. For example, while synonym replacement may be useful for general language-understanding models, it might not be suitable for domain-specific models where terminology precision is critical.

- **Balancing augmented data**: It's essential to ensure that the augmented data does not introduce its own biases or imbalances. The augmented instances should be mixed carefully with the original data to maintain a balanced and representative dataset.

- **Quality control**: After augmentation, the quality of the new data should be assessed to ensure that it is suitable for training. Poor-quality augmented data can be detrimental to the training process.

Data augmentation not only helps prevent overfitting by effectively increasing the size of the training set but also introduces the model to a wider range of linguistic phenomena, which is particularly important for tasks requiring high generalization capabilities.

Preprocessing

Preprocessing is a critical stage in preparing data for training LLMs. It involves various techniques to standardize and simplify the data, which can facilitate the model's learning process by reducing the complexity of the input space. Here's an expanded explanation of these preprocessing techniques:

- **Lowercasing**: This process converts all letters in the text to lowercase. It's a way to normalize words so that "The," "the," and "THE" are all treated as the same token, reducing vocabulary size. However, it may not always be appropriate, especially when case is significant, such as in proper nouns or in languages where case changes can alter the meaning of a word.

- **Stemming**: Stemming reduces words to their base or root form. For example, "running," "runs," and "ran" might all be stemmed to "run." This can help in consolidating different forms of a word, allowing the model to learn a more generalized representation. Stemming algorithms, however, can be too crude at times, as they often apply a set of rules without understanding the context (for example, "university" and "universe" might be incorrectly stemmed to the same root).

- **Lemmatization**: More sophisticated than stemming, lemmatization involves reducing words to their canonical or dictionary form (lemma). A lemmatizer takes into account the word's part of speech and its meaning in the sentence. Thus, "better" would be lemmatized to "good" when used as an adjective. Lemmatization helps in accurately condensing the various inflected forms of a word, which can be particularly useful for languages with rich morphology.

- **Normalization**: Text normalization includes correcting misspellings, expanding contractions (for example, converting "can't" to "cannot"), and standardizing expressions. This step ensures that the model isn't learning from or perpetuating errors in the data.

- **Removing punctuation and special characters**: Non-alphanumeric characters can be stripped out if they're not useful for the model's task. However, in tasks such as sentiment analysis or machine translation, punctuation can carry significant meaning and should be retained.

- **Handling stop words**: Commonly occurring words (such as "and," "the," or "is") that may not add much semantic value to the model's understanding can be removed. However, for some LLMs, especially those aimed at understanding complete sentences or paragraphs, stop words can provide essential context and should be kept.

- **Tokenization**: As previously mentioned, tokenization is the process of splitting text into manageable pieces or tokens. It's a necessary preprocessing step that directly affects the model's vocabulary.

For LLMs that aim to grasp the finer nuances of language or to generate human-like text, it's often important to maintain the original casing and form of words. In such cases, preprocessing should be carefully balanced to avoid losing meaningful linguistic information. For example, in NER, maintaining the case is crucial for distinguishing between common nouns and proper nouns.

Preprocessing must be tailored to the specific requirements of the LLM and the nature of the task it will perform. It's a delicate balance between simplifying the data to aid in learning general patterns and retaining enough complexity to allow the model to make nuanced linguistic distinctions.

Validation split

The **validation split** is a critical part of the data preparation process for training ML models, including LLMs. This process involves dividing the complete dataset into the following three distinct subsets, where each set plays a different role in the development and evaluation of the model:

- **Training set**: This is the largest portion of the dataset and is used for the actual training of the model. The model learns to make predictions or generate text by finding patterns in this data. The training process involves adjusting the model's weights based on the error between its predictions and the actual outcomes.

- **Validation set**: The validation set is used to evaluate the model during the training process, but it is not used to directly train the model. After each **epoch** (a complete pass through the training set), the model's performance is tested on the validation set. This performance serves as an indicator of how well the model is generalizing to unseen data. The results from the validation set are used to tune the model's hyperparameters, such as the learning rate, the model architecture, and regularization parameters. It can also be used for early stopping, which is a form of regularization where training is halted once the model's performance on the validation set stops improving.

- **Test set**: This is a set of data that the model has never seen during training and is not used in the hyperparameter tuning process. It is kept aside and used only after the model has been fully trained and validated. The test set provides an unbiased evaluation of the final model's performance and its ability to generalize to new data. It is the best estimate of how the model will perform in the real world on unseen data.

The way the data is split can vary depending on the amount of data available and the nature of the task. A common split ratio is 70% for training, 15% for validation, and 15% for testing, but this can be adjusted as needed. For instance, in cases where data is scarce, cross-validation techniques might be used, where the validation set is rotated through different subsets of the data.

It's crucial that the distribution of data in the training, validation, and test sets reflects the true distribution of the real-world data the model will encounter. This means that all classes or categories of interest should be represented proportionally in each set. The process of splitting the data should also be random to avoid introducing any bias.

A well-constructed validation split ensures that the LLM can be effectively tuned and ultimately performs well on the task it was designed for, while a final evaluation on the test set provides confidence in the model's real-world applicability.

Feature engineering

Feature engineering is a process in ML where specific information is extracted or derived from raw data to improve a model's ability to learn. In the context of LLMs and **natural language processing** (**NLP**), feature engineering can be particularly important for tasks that require an understanding of the structure and meaning of the text. A detailed look at what this might entail is as follows:

- **Parsing text for syntactic features**: Syntactic parsing involves breaking down a sentence into its grammatical components, such as nouns, verbs, and phrases. This can help an LLM understand the grammatical structure of sentences, which is especially useful for tasks such as translation or part-of-speech tagging. Syntactic features can include parse trees, parts of speech, and grammatical relationships between words.

- **Word embeddings**: Words can be converted into numerical vectors, known as embeddings, that capture their semantic meaning. Techniques such as Word2Vec, GloVe, or fastText analyze the text corpus and produce a high-dimensional space where semantically similar words are closer together. For LLMs, these embeddings provide a dense, information-rich representation of the input text.

- **Character embeddings**: Similar to word embeddings, character embeddings represent individual characters in a vector space. This can be useful for understanding morphology and is beneficial for languages where word boundaries are not as clear.

- **N-gram features**: N-grams are continuous sequences of n items from a given sample of text. Creating features based on n-grams can capture the context around words and phrases, which can be valuable for models that need to understand local context.

- **Entity embeddings**: In tasks that involve named entities, creating embeddings for entities that encode additional information about them (such as their type or relationships to other entities) can improve the model's performance.

- **Semantic role labeling**: This is the process of assigning roles to words in a sentence, identifying what role each word plays in the conveyed action or state. Features derived from semantic role labeling can enhance the model's understanding of sentence meaning.

- **Dependency parsing features**: Features derived from the dependencies between words in a sentence can help in understanding the relational structure of the text, which can be crucial for tasks that require a deep understanding of sentence semantics.

- **Part-of-speech tags**: These tags are helpful features for many NLP tasks, as they provide the model with information about the grammatical category of each word.

- **Transformations and interactions**: For certain tasks, it may be beneficial to engineer features that represent interactions between different words or parts of the text, such as whether two entities occur in the same sentence or paragraph.

- **Domain-specific features**: For specialized tasks, it might be necessary to engineer features that are specific to a domain. For example, in legal documents, features might represent references to laws or precedents.

- **Sentiment scores**: For sentiment analysis tasks, features might include sentiment scores of sentences or phrases, which can be obtained from pre-trained sentiment analysis models or lexicons.

The process of feature engineering requires domain knowledge and an understanding of the model's architecture and capabilities. While deep learning models, particularly LLMs, are capable of automatically learning representations from raw data, manually engineered features can still provide a performance boost, especially in cases where the model needs to understand complex relationships or when training data is limited.

Balancing the dataset

Balancing a dataset is a key aspect of preparing data for training LLMs. The goal is to create a dataset that represents the variety of outputs the model will need to predict without overrepresenting any particular class, style, or genre. This is essential to avoid biases that could skew the model's predictions when applied in real-world situations. Let's go through an expanded explanation of dataset balancing:

- **Class balance**: In classification tasks, it's crucial to have an approximately equal number of examples for each class. If one class is overrepresented in the training data, the model might become biased toward predicting that class more frequently, regardless of the input. Balancing can be achieved by undersampling the overrepresented classes, oversampling the underrepresented classes, or synthesizing new data for underrepresented classes.

- **Genre and style diversity**: For LLMs expected to generate or understand text across various genres and styles, the training data should include a mix of literary, journalistic, conversational, and technical writing, among others. This diversity ensures the model does not become biased toward a specific writing style or genre, which can limit its effectiveness.

- **Topic and domain coverage**: Including a wide range of topics and domains helps prevent the model from developing topic-specific biases. For instance, a model trained primarily on sports articles might struggle to understand or generate text related to medical information.

- **Demographic representation**: In scenarios where the model interacts with users or generates user-facing content, it's important for the dataset to represent the demographic diversity of the target audience. This involves including text that reflects different age groups, cultural backgrounds, and dialects.

- **Time period representation**: Historical balance can prevent temporal biases. Older texts can teach the model about outdated language forms, while newer texts ensure it is up to date with contemporary usage, including slang and neologisms.

- **Mitigating implicit biases**: Even with balanced classes and diversity, datasets can contain implicit biases that are less obvious. These can include gender, racial, or ideological biases. Active measures may be needed to identify and mitigate these biases, such as using fairness metrics or bias detection tools.

- **Data augmentation for balance**: When it's not possible to collect more data for underrepresented classes or styles, data augmentation techniques can artificially create additional examples to improve balance.

- **Sampling strategies**: When creating training, validation, and test splits, ensure that each split maintains the overall balance of the full dataset. Stratified sampling is a technique that can help achieve this by dividing the dataset such that each split reflects the same class proportions as the entire dataset.

- **Use class weights**: In cases where balancing data through sampling or augmentation is challenging, class weights can be used during training to give more importance to underrepresented classes, thereby mitigating bias in model predictions.

- **Regular evaluation**: Continually evaluate the model on a balanced validation set to monitor for biases. If biases are detected, the training data may need to be rebalanced or additional de-biasing techniques may need to be applied.

Balancing a dataset is not always straightforward, especially when dealing with complex or nuanced attributes. It requires thoughtful analysis and sometimes creative solutions to ensure that the final trained model behaves fairly and effectively across a wide range of inputs.

Data format

The format in which data is stored and handled can significantly impact the efficiency and effectiveness of training LLMs. Proper data formatting ensures that the data can be easily accessed, processed, and fed into the model during training. Here's an elaboration on the common formats and considerations:

- **JavaScript Object Notation (JSON)**: JSON is a lightweight data-interchange format that is easy for humans to read and write and easy for machines to parse and generate. It is particularly useful for datasets that have a nested or hierarchical structure. For instance, an annotated dataset for NLP might store each sentence along with its annotations in a structured JSON format, which can then be easily processed and used for training.

- **Comma-separated values (CSVs)**: CSV files are a common format for storing tabular data. Each line of the file is a data record, with individual fields separated by commas. This format is ideal for datasets that can be represented in a table format, such as a collection of text samples with associated labels. CSV files can be easily manipulated and processed with standard data processing tools and libraries, such as pandas in Python.

- **Plain text files**: For some tasks, especially those involving large amounts of unstructured text, plain text files may be the most straightforward format. They are simple to create and can be processed by almost any programming environment. However, they lack the structure to represent complex relationships or annotations, which might be necessary for certain types of training.

- **TFRecord**: TensorFlow's TFRecord file format is an efficient way to store data for TensorFlow models. It is particularly useful for datasets that need to be streamed from disk during training, which can be too large to fit into memory.

- `pickle`: Python provides a module named `pickle` that can serialize and de-serialize Python objects, converting them to a byte stream and back. While convenient, `pickle` files are specific to Python and may not be suitable for long-term data storage or for environments that use multiple programming languages.

- **Hierarchical Data Format version 5 (HDF5)**: HDF5 is a file format and set of tools for managing complex data. It is designed for flexible and efficient I/O and high-volume and complex data. HDF5 can be a good choice for datasets that require multi-dimensional arrays, such as word embeddings.

- **Parquet**: Parquet is a columnar storage file format that is optimized for use with big data processing frameworks. It is efficient for both storage and performance, supporting advanced nested data structures.

When converting data to the format best suited for the model's training framework, consider the following:

- **Scalability**: The format should be able to handle the scale of the data, both in terms of the number of records and the complexity of each record.

- **Performance**: The I/O performance of the format can be critical, especially when dealing with large datasets. The chosen format should enable efficient read and write operations.

- **Compatibility**: The format must be compatible with the tools and frameworks being used for model training. It should align with the expected input structure of the training pipeline.

- **Maintainability**: The ease of use and the ability to modify the dataset if needed are important. Some formats are more human-readable and easier to manipulate than others.

- **Integrity**: The format should preserve the integrity of the data, without loss or corruption.

By thoroughly preparing datasets, you can significantly enhance the performance of LLMs and ensure they learn a wide variety of language patterns and nuances. This groundwork is key to developing models that can generalize well and perform consistently across different tasks and domains.

Setting up your training environment

Establishing a robust training environment for LLMs involves creating a setup where models can learn effectively from data and improve over time. The steps to create such an environment are discussed next.

Hardware infrastructure

For training LLMs, the **hardware infrastructure** is an essential foundation that ensures the training process is efficient and effective. Here's an in-depth look at the key components:

- **Graphics processing units (GPUs)**: GPUs are specialized hardware designed to handle parallel tasks efficiently, which makes them ideal for the matrix and vector computations required in deep learning. Modern LLMs often necessitate the use of high-end GPUs with a large number of cores and substantial onboard memory to handle the computation loads.

- **Tensor processing units (TPUs)**: TPUs are custom chips developed specifically for ML workloads. They are optimized for the operations used in neural network training, offering high throughput for both training and inference. TPUs can be particularly effective for training LLMs at scale due to their high computational efficiency and speed.

- **High-performance CPUs**: While GPUs and TPUs handle the bulk of model training, high-performance CPUs are also important. They manage the overall control flow, data preprocessing, and I/O operations that feed data into the GPUs/TPUs.

- **Memory**: Adequate RAM is necessary to load training datasets, particularly when preprocessing and tokenizing large corpora. Insufficient memory can lead to bottlenecks, as data will need to be swapped in and out of slower storage.

- **Storage**: Fast, reliable storage is crucial for storing the large datasets used to train LLMs, as well as for saving the models' parameters and checkpoints during training. **Solid state drives (SSDs)** are preferred over **hard disk drives (HDDs)** for faster read/write speeds, which can significantly reduce data loading times.

- **Fast I/O capabilities**: Efficient I/O operations are vital to ensure that the training process is not I/O bound. This includes having a fast data pipeline that can supply data to the GPUs/TPUs without causing them to idle.

- **Networking**: For distributed training across multiple machines or clusters, high-bandwidth and low-latency networking are important to efficiently communicate updates and synchronize the model's parameters.

- **Cooling and power**: High-performance computing generates significant heat, so adequate cooling systems are necessary to maintain hardware integrity and performance. Similarly, a stable and sufficient power supply is critical to support the operation of high-end GPUs and TPUs.

- **Scalability**: The infrastructure should be scalable, allowing for the addition of more GPUs or TPUs as the complexity of the model or the size of the dataset grows.

- **Reliability and redundancy**: Systems should be robust, with redundancies in place to handle hardware failures, which can be common when training large models over extended periods.

- **Cloud computing platforms**: Many organizations opt for cloud-based services that offer scalable compute resources on-demand. Providers such as AWS, Google Cloud Platform, and Microsoft Azure offer GPU and TPU instances that can be rented, which can be a cost-effective alternative to purchasing and maintaining physical hardware.

- **Software compatibility**: Ensure that the hardware is compatible with the software stack and ML frameworks you plan to use, such as TensorFlow or PyTorch, which may have specific requirements for optimal performance.

Investing in the right hardware infrastructure is crucial for the successful training of LLMs, as it can greatly affect the speed of experimentation, the scale of training, and, ultimately, the quality of the models produced.

Software and tools

Selecting the appropriate software and tools is essential for the development and training of LLMs. The software stack includes not just ML frameworks, but also utilities that support data processing, model versioning, and experiment tracking. Here's a detailed look at these components.

ML frameworks

ML frameworks are pivotal in developing and deploying advanced algorithms, with each offering distinct features and advantages for various applications in the field:

- **TensorFlow**: An open source framework developed by the Google Brain team, known for its flexibility and robustness in building and deploying ML models. It offers comprehensive libraries for various ML tasks and supports distributed training.

- **PyTorch**: Developed by Meta's AI at Meta (formerly Facebook's AI Research lab), PyTorch is favored for its dynamic computation graph and user-friendly interface, making it particularly well suited for the research and development of deep learning models.

- **Hugging Face's Transformers**: A library built on top of TensorFlow and PyTorch, providing pre-built transformers and models for natural language understanding and generation. It simplifies the process of implementing state-of-the-art LLMs.

Data processing tools

Data science tools are specialized libraries that support the manipulation, analysis, and processing of data across different formats and complexities:

- **pandas/NumPy**: These are Python libraries that offer data structures and operations for manipulating numerical tables and time series. They are instrumental in handling and preprocessing structured data.

- **Scikit-learn**: A Python library that provides simple and efficient tools for data mining and data analysis. It includes functions for preprocessing and feature extraction.

- **spaCy**: An open source software library for advanced NLP in Python, offering robust tools for text preprocessing.

Version control systems

Version control systems are critical tools in software and ML development, managing changes in code, data, and models effectively:

- **Git**: A distributed version control system used for tracking changes in source code during software development. It is essential for managing code changes, especially when collaborating with a team.

- **Data Version Control (DVC)**: An open source version control system for ML projects. It extends version control to include data and model weights, enabling better tracking of experiments.

Experiment tracking and management

Experiment tracking and management tools are essential for streamlining the ML development process, from tracking progress to optimizing and deploying models:

- **MLflow**: This open source tool streamlines the ML life cycle, supporting deployment, fostering consistent experimental reproducibility, and managing the workflow. It helps track and organize experiments and manage and deploy models.

- **Weights & Biases**: A tool for experiment tracking, model optimization, and dataset versioning. It provides a dashboard for visualizing training processes and comparing different runs.

Containerization and virtualization

Containerization and virtualization technologies, such as Docker and Kubernetes, are crucial for the consistent deployment and scalable management of applications across environments:

- **Docker**: Platform-as-a-service solutions offered in this suite provide software packaged in modular units, leveraging OS-level virtualization, called **containers**. It ensures that the software runs reliably when moved from one computing environment to another.

- **Kubernetes**: An open source system used for automating the deployment, scaling, and management of containerized applications, ideal for managing complex applications such as LLMs.

Integrated development environments (IDEs) and code editors

IDEs and code editors, such as Jupyter Notebook and VS Code, are essential for efficient code creation, testing, and maintenance:

- **Jupyter Notebook**: A web-based open source application that enables the creation and distribution of documents with live code, equations, visualizations, and explanatory text
- **VS Code**: A source code editor that includes support for debugging, embedded Git control, syntax highlighting, and intelligent code completion

Deployment and monitoring

Tools such as TensorBoard and Grafana are pivotal for visualizing and monitoring ML models and systems:

- **TensorBoard**: With regard to deployment, this is a tool that offers key metrics and visualizations for ML workflows, supporting experiment tracking, model graph visualization, and more.
- **Grafana**: An open source platform for monitoring and observability. It can be used to create dashboards and alerts for your ML infrastructure.

Choosing the right set of software and tools depends on the specific requirements of the project, the team's expertise, and the existing infrastructure. It's important to select tools that integrate well with each other, have strong community support, and can scale with the project's needs.

Other items

In ML workflows, a variety of components beyond model building are critical for success, encompassing data handling to post-deployment operations and ethics:

- **Data pipeline**: Develop a scalable and automated data pipeline. This should include stages for data ingestion, preprocessing, transformation, augmentation, and feeding data into the training loop in batches.
- **Monitoring and logging**: Implement a system for monitoring and logging model performance and system health. Tools such as TensorBoard, Weights & Biases, or MLflow can track metrics, visualize training progress, and log experiments.
- **Hyperparameter tuning**: Use hyperparameter optimization tools to fine-tune the model's performance. Techniques such as grid search, random search, Bayesian optimization, or evolutionary algorithms can be employed to find the optimal set of hyperparameters.

- **Distributed training**: For very large models, consider setting up distributed training across multiple machines. This involves splitting the data and computation across different nodes to speed up the training process.

- **Regularization strategies**: Incorporate regularization strategies such as dropout, weight decay, or data augmentation to prevent overfitting and promote generalization in the model.

- **Testing and validation**: Create a robust testing and validation setup to evaluate the model against unseen data. This helps ensure the model's performance generalizes beyond the training data.

- **Security measures**: Implement security measures to protect data privacy and model integrity, particularly if working with sensitive information. This includes access controls, encryption, and compliance with data protection regulations.

- **Continuous integration / continuous deployment (CI/CD)**: Establish CI/CD pipelines for models to streamline updates and deployment. Automated testing and deployment can greatly enhance the efficiency of bringing model improvements to production.

- **Reproducibility**: Ensure that every aspect of the training process is reproducible. This includes using fixed seeds for random number generators and maintaining detailed versioning of datasets and model configurations.

- **Collaboration**: Facilitate collaboration among team members with tools that support versioning and sharing of models, data, and experiment results.

- **Documentation**: Keep comprehensive documentation for every aspect of the training environment. This should cover data preprocessing steps, model architectures, training procedures, and any assumptions or decisions made during the development process.

- **Ethical considerations**: Address ethical considerations proactively by reviewing datasets for potential biases, ensuring model transparency, and adhering to AI ethics guidelines.

By paying attention to these components, you can create a robust training environment that supports the development of effective LLMs capable of performing a wide range of tasks while maintaining high standards of quality and reliability.

Hyperparameter tuning – finding the sweet spot

Tuning hyperparameters is an important step in optimizing the performance of ML models, including LLMs. Let's look at a systematic approach to hyperparameter tuning:

- **Understand the hyperparameters**: Begin by understanding the hyperparameters that influence model performance. In LLMs, these can include learning rate, batch size, number of layers, number of attention heads, dropout rate, and activation functions, among others. The choice of values for these hyperparameters can affect the balance between memory requirements and training efficiency.

- **Establish a baseline**: Start with a set of default hyperparameters to establish a baseline performance. This can either come from the literature, default settings in popular frameworks, or empirical guesses.

- **Manual tuning**: Initially, perform some manual tuning based on intuition and experience to see how different hyperparameters affect performance. This can help set the bounds for more automated and systematic approaches.

- **Automated hyperparameter optimization**: Employ automated methods such as grid search, random search, or Bayesian optimization.

- **Grid search**: This exhaustively tries all combinations within a specified subset of the hyperparameter space.

- **Random search**: This samples hyperparameter combinations randomly instead of exhaustively. It's usually more efficient than grid search.

- **Bayesian optimization**: This uses a probabilistic model to predict the performance of hyperparameter combinations and chooses new hyperparameters to test by optimizing the expected performance.

- **Use gradient-based optimization**: For some hyperparameters, such as learning rates, gradient-based optimization methods can be applied. Learning rate schedulers can adjust the learning rate during training to help the model converge more effectively.

- **Model-based methods**: Techniques such as Hyperband and Bayesian optimization with Gaussian processes can be used to find good hyperparameters in fewer experiments by building a model of the hyperparameter space.

- **Early stopping**: Use early stopping during training to halt the process if the validation performance stops improving. This can also prevent overfitting.

- **Parallelize experiments**: If resources permit, run multiple sets of hyperparameters in parallel to speed up the search process.

- **Keep track of experiments**: Use experiment tracking tools to log hyperparameter values and corresponding model performance. This data is invaluable for understanding the hyperparameter space and can inform future tuning.

- **Evaluate on validation set**: Always evaluate the impact of hyperparameters on a held-out validation set to ensure that performance improvements generalize beyond the training data.

- **Prune unpromising trials**: Implement pruning strategies to stop training runs that don't show promise early on, saving computational resources.

- **Sensitivity analysis**: Perform a sensitivity analysis to understand which hyperparameters have the most significant impact on performance. Focus fine-tuning efforts on these parameters.

- **Final testing**: Once optimal hyperparameters are found, evaluate the model's performance on a test set to ensure that the improvements hold on unseen data.

- **Iterative refinement**: Hyperparameter tuning is often an iterative process. You may need to revisit steps based on test results or additional insights.

By methodically adjusting and evaluating the impact of different hyperparameters, you can optimize your LLM's performance for a variety of tasks and datasets. This process is part art and part science, requiring both systematic exploration and an intuitive understanding of model behavior.

Challenges in training LLMs – overfitting, underfitting, and more

Training LLMs presents several challenges that can affect the quality and applicability of the resulting models. Overfitting and underfitting are two primary concerns, along with several others.

Overfitting occurs when an LLM learns the training data too well, including its noise and outliers. This typically happens when the model is too complex relative to the simplicity of the data or when it has been trained for too long. An overfitted model performs well on its training data but poorly on new, unseen data because it fails to generalize the underlying patterns appropriately. To combat overfitting, techniques such as introducing dropout layers, applying regularization, and using early stopping during training are employed. Data augmentation and ensuring a large and diverse training set can also prevent the model from learning the training data too closely.

Underfitting is the opposite problem, where the model is too simple to capture the complexity of the data or has not been trained enough. An underfitted model performs poorly even on the training data because it doesn't learn the necessary patterns in the data. Addressing underfitting might involve increasing the model complexity, extending the training time, or providing more feature-rich data.

Other challenges in training LLMs include the following:

- **Data quality and quantity**: LLMs require vast amounts of high-quality, diverse data to learn effectively. Curating such datasets can be challenging and resource-intensive.

- **Bias in data**: The data used to train LLMs can contain biases, which the model will inevitably learn and replicate in its predictions. Efforts must be made to identify and mitigate biases in training datasets.

- **Computational resources**: Training LLMs demands substantial computational resources, which can be expensive and energy-intensive, posing scalability and environmental concerns.

- **Hyperparameter tuning**: Finding the optimal set of hyperparameters for an LLM is a complex and often time-consuming process. It requires extensive experimentation and can significantly affect model performance.

- **Interpretability**: LLMs, especially deep neural networks, are often considered "black boxes" because their decision-making processes are not easily understandable by humans. This lack of interpretability can be problematic, especially in applications that require trust and accountability.

- **Adaptability and continual learning**: After an LLM is trained, it should ideally be able to adapt to new data or tasks without extensive retraining. Developing models that can continually learn and adapt over time is an active area of research.

- **Evaluation metrics**: Proper evaluation of LLMs goes beyond simple accuracy or loss metrics. It must consider the context, coherence, and relevancy of the model's outputs, which can be difficult to quantify.

- **Ethical and legal considerations**: Ensuring that the use of LLMs adheres to ethical standards and legal regulations, especially regarding data privacy and user rights, is crucial.

- **Maintenance**: Once deployed, LLMs require ongoing maintenance to stay current with language trends, which can be a challenge given the rapid evolution of language and context in the real world.

Addressing these challenges requires a combination of technical strategies, careful planning, and adherence to ethical guidelines. As the field progresses, new techniques and methodologies are continually being developed to mitigate these issues and enhance the training and functionality of LLMs.

Summary

In this chapter, we laid out a comprehensive pathway for training LLMs, beginning with the imperative stage of data preparation and management. A robust corpus – varied, extensive, and balanced – is the bedrock upon which LLMs stand, requiring a diverse spectrum of text encompassing a broad scope of topics, cultural and linguistic representations, and temporal spans. To this end, we detailed the significance of collecting data that ensures a balanced representation and mitigates biases, hence fostering models that deliver a refined understanding of language.

Following the collection, rigorous processes of cleaning, tokenization, and annotation come into play to refine the quality and utility of data. These steps remove noise and standardize the text, breaking it into tokens that the model can efficiently process and annotate to provide contextual richness.

Data augmentation and preprocessing practices were emphasized as pivotal in expanding the scope of the data and standardizing it, thereby enabling the model to learn from a broader spectrum and prevent overfitting. The validation split underpinned the model's tuning process, ensuring its performance is robust, not just on the training set, but also on novel, unseen data.

Feature engineering was underscored as a critical step to extract and harness additional meaningful attributes from the data, enriching the model's understanding of language intricacies. This, along with the crucial step of balancing the dataset, ensures that the model's performance remains equitable across diverse inputs.

Proper data formatting was noted for setting the stage for efficient training and iteration, while the establishment of a solid training environment – with robust hardware and software infrastructure – was shown to be imperative for the successful training of LLMs. Hyperparameter tuning was addressed as a nuanced art and science necessary for optimizing the model's performance.

In conclusion, this chapter served as an extensive manual for practitioners in the field, presenting a well-orchestrated methodology for training LLMs that are capable, equitable, and adept at understanding and generating human language. It underlined the need for these models to function effectively, ethically, and responsibly across various applications.

In the next chapter, we will embark on explaining advanced training strategies so that you can achieve your desired objectives for your LLM applications.

Advanced Training Strategies

Expanding on the training strategy basics that were covered in the previous chapter, we'll delve into more sophisticated training strategies that can significantly enhance the performance of LLMs. We'll cover the subtleties of transfer learning, the strategic advantages of curriculum learning, and the future-focused approaches of multitasking and continual learning. Each concept will be solidified with a case study, providing real-world context and applications.

In this chapter, we're going to cover the following main topics:

- Transfer learning and fine-tuning in practice
- Curriculum learning – teaching LLMs effectively
- Multitasking and continual learning models
- Case study – training an LLM for a specialized domain

By the end of this chapter, you should understand the fundamental techniques that can be used to advance training strategies that boost the performance of LLMs.

Transfer learning and fine-tuning in practice

Transfer learning and fine-tuning are powerful techniques in the field of ML, particularly within NLP, to enhance the performance of models on specific tasks. This section will provide a detailed explanation of these concepts in practice.

Transfer learning

Transfer learning is the process of taking a pre-trained model that's been trained on a large dataset (often a general one) and adapting it to a new, typically related task. The idea is to leverage the knowledge the model has already acquired, such as understanding language structures or recognizing objects in images, and apply it to a new problem with less data available. In NLP, transfer learning has revolutionized the way models are developed. Previously, most NLP tasks required a model to be built from scratch, a process that involved extensive data collection and training time. With transfer learning, you can take a pre-trained model and adapt it to a new task with relatively little data.

Key benefits of transfer learning

Transfer learning has several benefits:

- **Computational efficiency**: Efficient computational strategies enhance ML processes by doing the following:

  - Reducing training time as you don't have to start from scratch

  - Lowering power consumption as you're fine-tuning a model rather than training a new one

- **Data efficiency**: Transfer learning boosts data efficiency in the following ways:

  - Requiring less labeled data

  - Effectively utilizing unlabeled data

- **Improved performance**: Transfer learning enhances model performance by offering the following:

  - Higher baseline accuracy

  - Better generalization on a new task due to the broader knowledge already acquired

- **Broad applicability**: Transfer learning showcases broad applicability by allowing the following:

  - Versatility across domains

  - Domain adaptation with minimal effort

- **Accessibility**: Transfer learning advances the accessibility of AI because of its ability to do the following:

  - Democratize AI by reducing the need for large datasets and extensive computing power

  - Enable rapid prototyping

Implementation considerations

The successful implementation of transfer learning hinges on several factors:

- The pre-trained model should be relevant to the new task
- Deciding how much of the pre-trained model to freeze and how much to fine-tune is crucial for the success of transfer learning
- The more similar the new task's data is to the data used in the pre-trained model, the more likely the transfer learning will be successful

Challenges and considerations

Now, let's learn how to navigate the challenges of transfer learning:

- If the new task's domain is very different from the text the model was originally trained on, the model might need significant adaptation.
- Finding the right fine-tuning approach can be complex. It requires carefully tuning the learning rate, deciding how many layers to fine-tune, and so on.
- Despite their efficiency, fine-tuning large models such as BERT or GPT still requires significant computational power, especially when dealing with large datasets or many fine-tuning iterations.
- There's a risk of overfitting to the new task if the fine-tuning process isn't managed carefully, especially with smaller datasets.
- Acquiring labeled data for fine-tuning can be costly and time-consuming, impacting overall efficiency and feasibility.
- Not all pre-trained models transfer knowledge effectively across different tasks, and identifying which models will work best can be challenging.
- Pre-trained models may carry biases from their original training data. This can be transferred to the new task if it's not mitigated properly.
- Understanding how and why a transfer learning model makes decisions can be difficult, particularly when using complex models such as deep neural networks.

Applications of transfer learning in NLP

Here are some of the applications of transfer learning in NLP:

- **Sentiment analysis**: Transfer learning tailors models to determine whether the sentiment of a piece of text is positive, negative, or neutral.

  A pre-trained model such as BERT can be fine-tuned with a smaller set of labeled sentiment data so that it specializes in understanding sentiments expressed in text, making it adept at classifying product reviews, social media posts, and so on. Fine-tuning is an integral part of transfer learning, where a pre-trained model is further trained on a specific dataset to adapt it for a particular task. This enables the model to use its existing knowledge to perform well on new tasks with limited data.

- **Question-answering**: In NLP, question-answering has been revolutionized by models on datasets by providing answers to questions based on a given context.

  BERT and GPT models, after being fine-tuned on datasets such as the **Stanford Question Answering Dataset** (**SQuAD**), can be proficient at reading a passage of text and answering questions about it, which is valuable for building conversational agents and search engines.

- **Language translation**: GPT and T5 models excel at the following aspects:

  - Translating text from one language into another

  - Fine-tuning models such as GPT and T5 on parallel corpora (text that's aligned in two or more languages) to perform translation tasks, reducing the need for extensive bilingual datasets for every language pair

- **Other tasks**: AI excels in the following areas:

  - Categorizing text into predefined categories

  - Identifying and classifying key elements in text into predefined categories such as the names of people, organizations, locations, and so on

  - Generating a concise and fluent summary of a long piece of text

Transfer learning is a powerful strategy in the ML toolkit that addresses key challenges in developing AI systems, especially when facing limitations regarding data, time, and computational capacity. Its benefits are most pronounced in scenarios where labeled data is scarce and the computational cost of training models from scratch is prohibitive. This approach not only streamlines the development process but also opens up the potential for innovation and application across a wide range of tasks and domains.

Fine-tuning

Let's take a closer look at fine-tuning:

- **Process**: Fine-tuning involves taking a pre-trained model and continuing the training process with a smaller, task-specific dataset. During fine-tuning, the model's weights are adjusted to better perform on the new task. This process is usually much faster than the initial training phase as the model has already learned a significant amount of general knowledge.

- **Customization**: Fine-tuning allows models to be customized to specific domains or applications. For instance, a model pre-trained on general English can be fine-tuned with legal documents to create a legal language model.

- **Challenges**: A potential challenge in fine-tuning is overfitting, where the model becomes too specialized to the fine-tuning dataset and loses its ability to generalize to new data. Careful monitoring, regularization techniques, and validation with a separate dataset are essential to avoid this.

Practical implementation of transfer learning and fine-tuning

In practice, transfer learning involves the following:

- **Selecting a pre-trained model:** The first step is to choose an appropriate pre-trained model. This choice depends on the nature of the task and the availability of pre-trained models suitable for the language or domain of interest.

- **Preparing task-specific data**: The data for fine-tuning should be closely related to the target task and properly labeled if necessary. It's also important to ensure the quality and diversity of this dataset to promote good generalization.

- **Model adaptation**: Adapting the model often involves adding or modifying the final layers so that the output is suitable for the specific task, such as changing the output to a different number of classes for classification tasks.

- **Hyperparameter tuning**: Adjusting hyperparameters such as the learning rate, batch size, and the number of epochs is crucial for effective fine-tuning. A lower learning rate is commonly used to make smaller, more precise adjustments to the weights.

- **Evaluation and iteration**: After fine-tuning, the model is evaluated using performance metrics relevant to the task. Based on these results, further iterations of fine-tuning may be performed to refine the model's performance.

In practice, transfer learning and fine-tuning have become standard procedures for developing NLP systems due to their efficiency and effectiveness. By building upon the vast knowledge that's acquired during pre-training, these techniques allow for the rapid development of specialized models capable of high performance on a wide array of NLP tasks.

Case study – enhancing clinical diagnosis with transfer learning and fine-tuning in NLP

Now, let's look at a hypothetical case study that focuses on transfer learning and fine-tuning in the healthcare industry.

Background

The healthcare industry continually seeks advancements in clinical diagnosis accuracy. With the advent of NLP, there is potential to automate and enhance the accuracy of diagnostic processes by analyzing patient records, clinical notes, and medical literature. In a hypothetical case study, a leading healthcare AI company embarked on a project to develop an NLP model that could support clinicians by providing more accurate diagnostic suggestions based on unstructured text data.

Challenge

The primary challenge was the sensitive nature of medical data, which is not only scarce but also heavily guarded due to privacy concerns. Furthermore, the company faced the daunting task of developing a model capable of understanding complex medical jargon and extracting relevant information from a variety of text styles and structures in patient records.

Solution – transfer learning and fine-tuning

To address these challenges, the company utilized transfer learning and fine-tuning methodologies by implementing the following phases:

- **Phase 1 – transfer learning implementation**:

  - **Model selection**: The company selected a pre-trained BERT model that had been trained on a broad range of general English text corpora

  - **Initial adaptation**: They adapted the model to the medical domain using a large-scale medical dataset, including publications and anonymized patient notes, to grasp the medical lexicon and sentence structures

- **Phase 2 – fine-tuning the model**:

  - **Data preparation**: A smaller, highly specialized dataset was curated, consisting of annotated clinical notes and diagnosis records that represented a wide spectrum of cases

  - **Model training**: The pre-trained BERT model was fine-tuned with this dataset, focusing on disease markers and diagnostic patterns

  - **Validation and testing**: The model was rigorously validated against a control set that was reviewed by medical professionals to ensure accuracy and reliability

Results

The fine-tuned NLP model demonstrated a remarkable improvement in identifying diagnostic entities and suggesting accurate diagnoses from clinical notes. It showed the following:

- A 20% increase in diagnosis accuracy compared to the baseline model
- A significant reduction in false positives, which is crucial for medical applications
- Improved efficiency, reducing the time taken for preliminary diagnosis

Impact

The implementation of transfer learning and fine-tuning resulted in several impactful outcomes:

- **Support for clinicians**: The model became an invaluable tool for clinicians, providing them with quick, accurate diagnostic suggestions
- **Resource optimization**: It reduced the time clinicians spent on preliminary diagnosis, allowing them to focus on patient care
- **Scalability**: The approach demonstrated a scalable model for incorporating AI in healthcare, opening pathways for further innovations

Conclusion

This case study illustrates the practical benefits of transfer learning and fine-tuning in NLP within the healthcare sector. By leveraging these techniques, the company was able to create a tool that enhanced the accuracy of clinical diagnoses. This project not only exemplifies the effectiveness of these methodologies in dealing with domain-specific challenges but also sets a precedent for future AI-driven healthcare solutions.

Now, let's delve into how LLMs are taught effectively.

Curriculum learning – teaching LLMs effectively

Curriculum learning is an approach in ML, particularly when training LLMs, that mimics the way humans learn progressively from easier to more complex concepts. The idea is to start with simpler tasks or simpler forms of data and gradually increase the complexity as the model's performance improves. This approach can lead to more effective learning outcomes and can help the model to better generalize from the training data to real-world tasks. Let's take a closer look at this approach.

Key concepts in curriculum learning

Here, we'll review some key concepts in curriculum learning that you should be aware of.

Sequencing

Sequencing in curriculum learning is analogous to the educational curricula in human learning, where subjects are taught in a logical progression from simple to complex. In ML, the following are applicable:

- **Graduated complexity**: Training begins with easier instances to give the model a foundational understanding before it tackles more complex scenarios

- **Task decomposition**: Complex tasks are broken down into simpler, more manageable subtasks that are learned in sequence

- **Sample selection**: Initially, samples that are more representative of the general distribution or are less noisy are chosen to help the model learn the basic patterns before outliers or edge cases are introduced

In NLP, sequencing might involve starting with basic vocabulary and grammar before introducing complex sentences, metaphors, or domain-specific jargon. For example, a language model might be exposed to simple sentences ("The cat sat on the mat") before encountering complex ones ("Despite the cacophony, the cat, undisturbed, sat on the checkered mat").

Pacing

Pacing is about controlling the speed at which new concepts are introduced:

- **Adaptive learning rate**: Adjusting the pace of learning based on the model's performance, similar to a teacher providing feedback to a student

- **Performance thresholds**: Moving to more complex materials only after the model achieves a certain level of performance on the current material

- **Staged difficulty**: Introducing new difficulty levels in stages, with each stage having a set of criteria for mastery before progression

In the context of curriculum learning, pacing ensures that the model has sufficiently learned from current examples before moving on to more challenging ones. This could be akin to ensuring a student understands basic algebra before introducing them to calculus.

Focus areas

The concept of focus areas in curriculum learning relates to concentrating on particular aspects of the learning task at different stages of the training process:

- **Concept isolation**: This involves teaching specific concepts in isolation before integrating them with other learned concepts. For example, in language learning, this could involve focusing on the present tense before introducing past or future tenses.

- **Attention shifting**: This involves shifting the model's focus during training to various aspects of the data. In NLP, a model might focus on syntax first before shifting focus to semantic analysis.

- **Progressive refinement**: This involves starting with a broad approximation of the target function and then refining the model's understanding over time. This is akin to teaching broad strokes in art before focusing on the finer details.

For instance, in language models, initial focus areas may include basic sentence structure and vocabulary, before more complex linguistic features, such as irony or ambiguity, are considered.

Benefits of curriculum learning

Curriculum learning provides the following benefits:

- **Efficiency**: Efficiency in AI training is achieved through the following:

  - **Accelerated initial learning**: By beginning with simpler tasks, the AI model can quickly achieve initial success, which can reinforce the correct learning patterns and boost its learning curve.

  - **Resource optimization**: Curriculum learning can lead to more efficient use of computational resources. Training on simpler tasks first generally requires less computational power, and as the model's capability increases, so can the computational investment.

  - **Reduced training time**: As the model is not immediately overwhelmed with complex tasks, it can converge to a good solution faster, making the overall training process more time-efficient.

- **Performance**: Curriculum learning provides various benefits:

  - **Improved accuracy**: Models trained using a curriculum tend to develop a more nuanced understanding of the data, leading to better accuracy and performance on their tasks

  - **Stronger foundational knowledge**: The model builds a robust foundation of the basics, which is essential for understanding more intricate patterns and structures later on

  - **Less prone to overfitting**: With a focus on general principles first, models are less likely to overfit to the noise in more complex training examples

- **Generalization**: Generalization is enhanced through the following aspects:

  - **Better transferability**: A model that has a strong base in fundamental concepts may be more capable of transferring what it has learned to new, unseen data, which is crucial for real-world applications

  - **Adaptability to variations**: Staged learning helps the model adapt to variations within the data, leading to better performance on tasks that were not part of the training set

  - **Handling of real-world complexity**: By gradually introducing complexity, the model can better mimic the progression of learning required to handle complex real-world tasks

- **Improved interpretability**: Curriculum learning enhances interpretability in the following ways:

  - **Providing a clearer understanding of model behavior**: Curriculum learning provides insights into how models develop their understanding over time, making their decision-making processes more interpretable.

  - **Facilitated debugging and analysis**: By following a structured learning path, it becomes easier to identify and address errors. This is because the model's learning stages are clearer and more logical.

Additional considerations

The following are some additional considerations regarding curriculum design:

- **Curriculum design**: The design of the learning curriculum must be thoughtful and strategic to ensure that the model is not only learning efficiently but also developing the capacity to handle the complexity of real-world applications

- **Balanced progression**: The progression from simple to complex needs to be balanced to ensure that the model is challenged just enough to learn without being overwhelmed or plateauing in its learning journey

- **Evaluation metrics**: It is crucial to have proper evaluation metrics in place to assess the effectiveness of the curriculum and the model's readiness to progress to more challenging tasks

Curriculum learning addresses some of the fundamental challenges in training LLMs by structuring the learning process in a more human-like fashion. By optimizing the order and complexity of training data, this approach not only makes the training process more efficient but also enhances the performance and generalization capabilities of the models. Such benefits are particularly important as LLMs are increasingly being deployed in diverse and complex real-world scenarios, where adaptability and robustness are key to success.

Implementing curriculum learning

Implementing curriculum learning in ML and AI involves several critical steps to ensure that the model can effectively progress from learning simple concepts to mastering complex ones. We'll take a closer look at these steps here.

Data organization

Organizing training data by complexity is the cornerstone of curriculum learning. This process can be quite nuanced, depending on the domain and the specific tasks the model is being prepared for. The following are key aspects that need to be addressed:

- **Complexity metrics**: Developing metrics to evaluate the complexity of data is essential. For language models, this might involve sentence length, vocabulary difficulty, or syntactic complexity. In other domains, complexity could be measured by the number of features, the ambiguity of labels, or the rarity of the data points.

- **Expert involvement**: Involving subject matter experts can be critical, especially when complexity metrics are not clear-cut or when the data requires domain-specific insight to be categorized properly.

- **Automated sorting**: ML techniques, such as clustering algorithms, can be used to sort data into complexity tiers automatically. These methods might use feature vectors to determine similarity and group data points accordingly.

Model monitoring

Continuous evaluation of the model's performance is necessary to gauge when it's ready to move on to more difficult material. This can be achieved by using the following:

- **Performance metrics**: Defining clear performance metrics such as accuracy, precision, recall, or a domain-specific metric is necessary to objectively assess the model's progress

- **Feedback loops**: Implementing feedback mechanisms that can guide the training process and inform decisions about when to introduce more complex data

- **Early stopping**: This technique can prevent overfitting on simpler data and prompt the transition to more complex stages when the model's improvement in the current stage diminishes

Dynamic adjustments

The ability to adapt the training process dynamically is a key feature of effective curriculum learning. This can be incorporated with the help of the following:

- **Adaptive pacing:** The curriculum should allow for changes in pacing based on real-time performance, slowing down when the model struggles and accelerating when it masters a concept quickly.

- **Curriculum refinement**: The initial curriculum might need to be refined as the model's learning patterns emerge. This could involve adding more intermediate steps or revising the complexity measures.

Task-specific curricula

Designing curricula that are tailored to the final tasks of the model can significantly enhance its effectiveness. For this purpose, you need to manage the following:

- **Task analysis**: A thorough analysis of the end tasks can help you identify the core skills and knowledge the model needs to acquire. For example, customer service models need to understand colloquial language and empathy, while medical models must interpret clinical terminology accurately.

- **Curriculum design**: The curriculum should reflect the progression of skills and knowledge required for the model to perform its final tasks. For instance, a curriculum for a medical diagnosis model might start with general medical knowledge before focusing on symptoms and treatments for specific conditions.

Implementing curriculum learning is a complex process that requires careful planning, continuous monitoring, and the flexibility to adapt the curriculum as the model learns. It's a strategic approach that, when executed well, can significantly improve the efficiency and effectiveness of AI models, particularly in specialized or complex domains. By tailoring the learning process to the model's needs and the intricacies of the task at hand, curriculum learning can lead to AI systems that are not only highly competent in their designated tasks but also capable of generalizing their knowledge to new, related challenges.

Challenges in curriculum learning

Curriculum learning comes with its own set of challenges. Let's take a look.

Defining complexity

Determining the complexity within training data is a critical and non-trivial aspect of curriculum learning. In the context of language, this is particularly challenging due to the following aspects:

- **The multidimensional nature of language**: Language complexity is not one-dimensional; it includes syntactic complexity, semantic richness, pragmatics, and more. An example that is simple in that one respect might be complex in another.

- **Subjectivity**: What one model or domain expert considers complex, another might not. This subjectivity can make standardizing a measure of complexity difficult.

- **Automated complexity measures**: Developing automated measures that accurately reflect complexity requires advanced algorithms that can potentially incorporate linguistic, contextual, and domain-specific features.

Curriculum design

Creating an effective curriculum is akin to developing an educational course for a human student – it requires understanding how the "student" (in this case, the model) learns about the following:

- **Domain expertise**: The designer of the curriculum needs to have a thorough understanding of the domain to ensure that all the necessary concepts are taught in an appropriate sequence.

- **Model understanding**: Different models may learn in different ways. Understanding the learning dynamics of the specific model being used is crucial for designing an effective curriculum.

- **Iterative process**: Designing a curriculum is not a one-time task; it often requires iterations and modifications as the model's performance on the tasks is observed and analyzed.

Balancing breadth and depth

Striking the right balance between a broad understanding and deep expertise is a delicate task that includes various aspects:

- **Breadth**: Ensuring the model has a comprehensive understanding of a wide range of topics or skills is important for generalization. However, too much breadth can lead to a superficial understanding of each topic.

- **Depth**: Providing in-depth knowledge in certain areas is necessary for expertise. However, focusing too deeply on one area can limit the model's ability to handle a variety of tasks.

- **Practical application**: The ultimate goal is to deploy the model in real-world applications. Therefore, the curriculum should focus on achieving the right mix of breadth and depth to prepare the model for the tasks it will encounter.

Generalization and overfitting

Managing generalization and overfitting is crucial in curriculum learning:

- **Generalization**: The curriculum must be designed to ensure that the model can generalize its learning to new and unseen data, which is often challenging when creating a staged learning process

- **Overfitting**: There is a risk of overfitting to simpler tasks if the curriculum does not progressively increase in complexity or if too much emphasis is placed on easy examples

Evaluation and metrics

Evaluating the effectiveness of curriculum learning requires the following careful considerations:

- **Choosing the right metrics**: Determining which metrics best reflect the model's progress and effectiveness at each stage of the curriculum can be challenging

- **Continuous monitoring**: Regularly evaluating model performance to adjust the curriculum requires significant resources and ongoing analysis

- **Benchmarking**: Establishing benchmarks to compare the effectiveness of different curriculum designs is essential but can be difficult due to variability in tasks and models

Model-specific challenges

Each model may present unique challenges when implementing curriculum learning:

- **Architecture-specific considerations**: Different models may require tailored curriculum designs that consider their specific architecture and learning dynamics

- **Resource constraints**: The computational and data requirements of different models can vary widely, influencing how the curriculum can be structured and executed

Practical strategies for addressing challenges

The following are some practical strategies that can be used:

- **Expert collaboration**: Working with domain experts can help in accurately defining complexity and designing a well-rounded curriculum

- **Incremental development**: Building the curriculum incrementally, starting with a basic structure and then refining it based on the model's performance, can make the process more manageable

- **Evaluation and feedback**: Regularly evaluating the model's performance and incorporating feedback can help in fine-tuning the curriculum to better meet the model's learning needs

- **Modular design**: Creating a modular curriculum that can be adjusted or reorganized easily allows for more dynamic learning paths tailored to the model's progression

Curriculum learning, while powerful, requires thoughtful implementation to overcome its inherent challenges. The intricacies of defining complexity, designing the curriculum, and achieving a balance of breadth and depth are substantial hurdles. However, with a careful approach that includes expert input, iterative design, and ongoing evaluation, these challenges can be navigated successfully. The outcome is a more effective training process that produces models capable of sophisticated understanding and performance.

Case study – curriculum learning in training LLMs for legal document analysis

This case study focuses on curriculum learning in the legal industry.

Background

In a hypothetical case study, a legal tech start-up aimed to develop an LLM capable of parsing and understanding complex legal documents to provide summaries and actionable insights. The goal was to assist lawyers by automating the preliminary review of case files, contracts, and legislation, which are typically dense and filled with specialized language.

Challenge

The main challenge was the complexity of legal language, which included a wide range of vocabulary, specific jargon, and intricate sentence structures. Traditional training methods proved inefficient as the model struggled with the advanced nuances of legal texts after being trained on general language data.

Solution – curriculum learning

To overcome this, the company implemented a curriculum learning approach, structuring the model's training to progressively increase in complexity, closely aligning with the cognitive steps a human expert would take when learning the legal domain. This involved the following phases:

- **Phase 1 – structured learning progression**:

  - **Simple to complex**: The LLM began by learning simple legal definitions and moved toward understanding complex contractual clauses

  - **Segmented learning**: Training was segmented into phases, starting with general legal principles before progressing to specifics such as tax law, intellectual property rights, and international regulations

- **Phase 2 – incremental complexity increase**:

  - **Controlled vocabulary expansion**: Vocabulary was introduced in a controlled manner, starting with general legal terms before more specialized terms were incorporated

  - **Complexity in context**: The model was exposed to increasingly complex sentences, starting from clear-cut case law to convoluted legal arguments

Results

The curriculum learning approach yielded a highly efficient LLM that showed the following:

- A 35% improvement in the comprehension of legal terminology compared to the baseline model trained without curriculum learning

- A 25% increase in accuracy when summarizing legal documents

- Enhanced ability to identify relevant legal precedents and citations

Impact

The successful implementation of curriculum learning significantly impacted the start-up's objectives:

- **Efficiency in legal reviews**: The LLM reduced the time lawyers spent on initial document reviews by automating the process of extracting key points

- **Scalability of legal services**: Smaller law firms, previously limited by resource constraints, could scale their operations by utilizing AI for routine document analysis

- **Consistency and reliability**: The LLM provided consistent and reliable analysis, reducing human error in initial reviews

Conclusion

This case study demonstrates the effectiveness of curriculum learning in training LLMs for specialized tasks. By mimicking the natural progression of human learning, the start-up was able to create a model that understood and analyzed legal documents with high accuracy. This approach not only proved to be a breakthrough in legal technology but also showcased a scalable method for applying AI in specialized fields, potentially transforming how professionals engage with dense and specialized texts.

Next, we'll delve into multitasking and continual learning models.

Multitasking and continual learning models

Multitasking and continual learning models represent two pivotal areas of research in the field of AI and ML, each addressing distinct but complementary challenges related to the flexibility and adaptability of AI systems.

Multitasking models

Multitasking models, also known as **multi-task learning** (**MTL**) models, are designed to handle multiple tasks simultaneously, leveraging the commonalities and differences across tasks to improve the performance of each task. This hypothesis is grounded in cognitive science, suggesting that human learning often involves transferring knowledge across different but related tasks. In AI, this translates into models that can process and learn from multiple tasks simultaneously, optimizing shared neural network parameters to benefit all tasks involved.

The central idea is to share representations between related tasks to avoid learning each task in isolation, which can be inefficient and require more data. This approach can lead to models that are more generalizable and efficient as they can learn useful features from one task that apply to others.

Key characteristics

The key characteristics of multitasking models are as follows:

- **Shared architectures**: Multitasking models are designed to handle multiple tasks that can benefit from shared representations. Here's how shared architectures work and their benefits:

  - **Layer sharing**: The initial layers of the network are shared among all tasks. These layers typically learn the basic patterns in the data that are common across tasks. For example, in a visual recognition model, these layers might detect edges and shapes that are fundamental to many different objects.

  - **Specialization in later layers**: As the network progresses, layers become more specialized for individual tasks. This can be seen as a divergence point where task-specific knowledge is refined and applied. In our visual recognition example, these specialized layers would learn patterns specific to different categories, such as animals, vehicles, or furniture.

  - **Efficient learning**: By sharing parameters, these architectures require fewer resources than when training separate models for each task as they don't need to relearn the same general features for each new task.

  - **Feature reuse**: Shared architectures can lead to feature reuse, where a feature learned for one task can be beneficial for another. This may not have been possible if the tasks were learned in isolation.

- **Joint learning**: Joint learning refers to the training on multiple tasks simultaneously. This approach has several advantages:

  - **Cross-task feature learning**: When models are trained jointly, they can learn features that are useful across multiple tasks, which might not be learned when tasks are trained independently

  - **Improved generalization**: Training on a diverse set of tasks can help the model generalize better to new tasks or data as it learns to extract and utilize broadly applicable features

- **Balanced learning**: Joint learning can help prevent the model from overfitting to one task by balancing the learning signals from multiple tasks

- **Regularization effect**: MTL inherently incorporates a form of regularization due to its training dynamics:

  - **Parameter sharing as regularization**: By sharing parameters across tasks, the model is implicitly regularized. This is because the shared parameters must be useful across all tasks they are shared among, preventing the model from overfitting to the idiosyncrasies of a single task's training data.

  - **Constraints from multiple tasks**: Training with multiple tasks imposes additional constraints on the model as it has to perform well on all tasks simultaneously. This can help reduce the model's capacity to memorize the training data and instead force it to find underlying patterns that are more generally applicable.

  - **Noise robustness**: Exposure to multiple tasks during training can also make the model more robust to noise as noise patterns are less likely to be consistent across different tasks, and hence, the model is less likely to learn them.

These characteristics make multitasking models particularly powerful for complex applications where multiple related problems need to be solved simultaneously, and where the benefits of shared knowledge, joint learning, and regularization effects can lead to more robust, generalizable, and efficient solutions.

Advanced techniques in MTL

Now, let's consider some advanced techniques in MTL:

- **Cross-stitch networks**: These are sophisticated versions of multitasking models that allow the optimal level of task sharing to be learned automatically. Unlike traditional shared architectures, cross-stitch units enable the network to learn how much information to share between tasks dynamically.

- **Task attention networks**: By incorporating attention mechanisms, multitasking models can weigh the importance of shared features differently for each task, allowing the model to focus more on relevant features for a given task while ignoring less useful information.

Applications

Multitasking models are widely used in various domains, including NLP, where a single model may perform entity recognition, sentiment analysis, and language translation. They are also prevalent in computer vision for tasks such as object detection, segmentation, and classification within the same framework. Let's review their applications.

NLP

In NLP, multitasking models are highly beneficial because many tasks share common linguistic features and structures. A single model that can capture these shared elements can be applied to multiple NLP tasks:

- **Entity recognition**: This task involves identifying and classifying key information in text into predefined categories, such as the names of people, organizations, and locations. Multitask models can learn contextual cues from sentence structures that help in recognizing entities.

- **Sentiment analysis**: This task involves understanding the sentiment expressed in a piece of text, whether it's positive, negative, or neutral. Models that have been trained to recognize sentiment can also benefit from, and contribute to, understanding language nuances that are required for other tasks, such as language translation.

- **Language translation**: Translation requires the model to understand the syntax and semantics of both the source and target languages. Multitasking models can leverage the deep understanding of language gained from other NLP tasks to improve translation accuracy.

The shared layers in a multitask model handle the common aspects of language, such as grammar and common vocabulary, while task-specific layers fine-tune the model's outputs for each task.

Computer vision

In computer vision, multitasking models take advantage of shared visual features across different tasks:

- **Object detection**: This involves locating objects within an image and classifying them. The initial layers of the multitasking model might learn to detect edges and textures, which are useful for many vision tasks.

- **Segmentation**: In image segmentation, the task is to assign a label to each pixel in an image so that pixels with the same label share certain characteristics. Multitasking models benefit from understanding general shapes and boundaries learned during object detection.

- **Classification**: Image classification involves assigning a class label to an image (or parts of an image). MTL can help with classification by leveraging feature detectors developed for detection and segmentation tasks.

In each of these tasks, the early layers of a multitasking model capture generic features such as shapes and edges, while later layers become more specialized, such as recognizing specific object features for detection or finer details for segmentation.

Advantages across domains

The use of multitasking models across these domains offers several advantages:

- **Resource efficiency**: Training one model for multiple tasks is more resource-efficient than training separate models for each task

- **Consistency**: Having a single model perform multiple related tasks can lead to consistency in the performance and integration of the model's outputs

- **Cross-task learning**: The model can leverage what it learns from one task to improve its performance on another, which is a form of inductive transfer that can improve overall learning efficiency

Challenges and solutions

MTL faces challenges:

- **Task interference**: A significant challenge in MTL is task interference, where the learning of one task negatively impacts the performance of another. Advanced regularization techniques and architecture designs, such as task-specific batch normalization and soft parameter sharing, are explored to mitigate this issue.

- **Optimizing task weighting**: Determining the right balance in learning between tasks remains a challenge. Adaptive weighting methods, which dynamically adjust the importance of each task's loss function during training, are being developed to address this.

Continual learning models

Continual learning models, also known as lifelong learning models, are designed to learn continuously from a stream of data, acquiring, retaining, and transferring knowledge across tasks over time. The primary challenge these models address is avoiding catastrophic forgetting, which occurs when a model learns a new task at the expense of previously learned tasks.

Key characteristics

Here are the key characteristics of continual learning models:

- **Knowledge retention**: The main goal of knowledge retention in continual learning models is to overcome what's known as catastrophic forgetting, which is the tendency of a neural network to completely forget old knowledge upon learning new information. Here's how knowledge retention is typically addressed:

  - **Replay mechanisms (experience replay)**: This technique involves storing data from previously learned tasks and reintroducing it into the learning process periodically.

This can prevent the model from forgetting previously learned information. This replay can be done by doing the following:

- Randomly sampling and reintegrating old data into new training batches

- Using a generative model to recreate the distribution of previous tasks and using this synthetic data for retraining

- Maintaining a subset of the original training data for old tasks, where it can be used alongside new task data during further training iterations

- **Regularization methods**: Regularization strategies are employed to protect the knowledge that the model has already acquired.

- **Elastic weight consolidation (EWC)**: This technique adds a penalty to the loss function based on how important each network parameter is to previously learned tasks. It effectively creates a constraint that discourages the model from changing important parameters when learning new information.

- **Synaptic intelligence**: Similar to EWC, synaptic intelligence estimates the importance of each synapse (connection between neurons) for the tasks learned so far and then penalizes changes to the most crucial synapses.

- **Knowledge distillation**: The model's knowledge is distilled and transferred during the training on new tasks. This is often by using the model's predictions as "soft targets" during further training.

- **Architectural approaches**: Some models incorporate architectural strategies to allow for knowledge retention.

- **Progressive neural networks**: These networks grow over time by adding new columns of neurons for new tasks while freezing the columns associated with previous tasks.

- **Dynamic network expansion**: Here, the model architecture is dynamically expanded to accommodate new knowledge, often by adding new neurons or layers when learning new tasks.

- **Complementary Learning Systems (CLS)**: This neuroscience-inspired approach involves having dual memory systems in the model – one for rapid learning and another for slow consolidation of knowledge – akin to the hippocampus and neocortex in the human brain.

Challenges and considerations

There are various challenges and considerations in continual learning:

- **Memory management**: Deciding how much old data to store for replay can be challenging, especially when considering constraints on computational resources and storage

- **Balancing stability and plasticity**: Models must balance the ability to retain old knowledge (stability) with the ability to learn new tasks (plasticity)

- **Task interference**: When tasks are very different, learning a new task might interfere with the performance of old tasks, even when using replay or regularization techniques

- **Data distribution shift**: Adapting to evolving data distribution requires continual learning models to do the following:

  - **Adapt to changes**: Continual learning models must adapt to shifts in data distribution over time, which can be challenging if the shifts are abrupt or unpredictable

  - **Be robust to variability**: Ensuring robustness to changes in data distribution is essential for maintaining model performance across different tasks and environments

- **Evaluation and benchmarking**: You must do the following to ensure an effective model evaluation and comparison process:

  - **Set appropriate benchmarks**: Establishing benchmarks that accurately reflect the model's continual learning capabilities across tasks is essential but challenging due to task variability

  - **Provide consistent evaluation metrics**: Using consistent and relevant metrics to evaluate performance over time is critical to assess the model's effectiveness in learning and retaining knowledge

  - **Perform comparative analysis**: Conducting comparative analysis with other models and techniques is necessary to understand the relative strengths and weaknesses of the approach

The capacity for knowledge retention in continual learning models is critical for their development and deployment in real-world applications, where they must adapt to new data and tasks over time. By employing strategies such as replay mechanisms, regularization methods, and architectural adjustments, these models aim to retain previously learned information while continually incorporating new knowledge. This area is still under active research, with many promising approaches being explored to tackle the inherent challenges of continual learning.

Applications

Continual learning is a transformative approach in AI with wide-ranging applications across different fields. By enabling models to learn incrementally and adapt to new data without forgetting previous knowledge, continual learning models can be applied to many real-world scenarios where adaptability and the ability to learn from new experiences are crucial. Here are some applications of continual learning:

- **Personalized recommendation systems**:

  - **Adapting to user behavior**: Continual learning allows recommendation systems to adapt to changing user preferences over time, which is essential as interests and behaviors evolve.

- **Dynamic content**: As new content is constantly being created, recommendation systems must continually learn to include these in their suggestions. Continual learning methods ensure that the system can integrate new content without losing its ability to recommend older but still relevant items.

- **Long-term user satisfaction**: By retaining knowledge of a user's historical preferences while adapting to their current interests, continual learning helps in maintaining long-term user satisfaction and engagement.

- **Autonomous robots**:

  - **Real-world interaction**: Robots operating in real-world environments encounter dynamic and unforeseen situations. Continual learning enables them to accumulate experience and improve their decision-making processes over time.

  - **Skill acquisition**: As robots are exposed to new tasks, they need to acquire new skills without forgetting the old ones. Continual learning models can help them integrate these new skills seamlessly.

  - **Environment adaptation**: For robots that navigate varying environments, the ability to learn from these new experiences and adjust their models accordingly is facilitated by continual learning techniques.

- **Healthcare monitoring**:

  - **Patient data analysis**: Continual learning can be applied to monitor patients' health over time, adjusting to new data such as changes in vital signs or the progression of a disease, to provide timely and personalized healthcare

  - **Adapting treatment plans**: As more patient data becomes available, healthcare models can use continual learning to adjust treatment plans based on the effectiveness of previous treatments and evolving health conditions

- **Financial markets**:

  - **Market trend analysis**: Financial models need to adapt to ever-changing market conditions. Continual learning allows these models to assimilate new market data continuously, helping to predict trends and make informed decisions.

  - **Risk management**: Continual learning helps in adjusting risk models in finance as new financial instruments are introduced and market dynamics change.

- **Automotive systems**:

  - **Self-driving cars**: Continual learning is key for self-driving car algorithms that must adapt to diverse and changing driving conditions, traffic patterns, and pedestrian behaviors

- **Online services**:

  - **Content moderation**: Services that rely on content moderation must constantly update their models to understand new slang, symbols, and changing contexts. Continual learning enables these systems to evolve with the language and societal norms.

- **Education**:

  - **Adaptive learning platforms**: Educational platforms can use continual learning to personalize the learning experience, adapting to the changing proficiency and learning speed of each student

- **Software applications**:

  - **User interface adaptation**: Software applications can use continual learning to adapt their interfaces based on user behavior patterns, creating a more personalized and efficient user experience

In each of these applications, the primary advantage of continual learning is its dynamic adaptability, which ensures that models remain relevant and effective over time as they encounter new information. This adaptability is particularly important in our rapidly changing world, where static models can quickly become outdated. Continual learning represents a significant step toward more intelligent, responsive, and personalized AI systems.

Challenges and solutions

Continual learning has various challenges:

- **Memory overhead**: As models learn new tasks, the requirement for memory and computational resources can grow substantially. Methods such as dynamic network expansion, which selectively grows the model's capacity, and memory-efficient experience replay techniques are under development to manage this growth sustainably.

- **Balancing stability and plasticity**: Continual learning models must balance the need to retain previously learned information (stability) with the need to adapt to new information (plasticity). Techniques such as synaptic intelligence, which measures and protects the contribution of synapses to task performance, aim to strike this balance effectively.

Synergistic potential and future horizons

The integration of multitasking and continual learning models holds the promise of creating AI systems that are not only versatile across a broad range of tasks but also adaptive to new challenges over time. This synergy could lead to the development of AI that is more akin to human intelligence and capable of lifelong learning and adaptation.

Emerging research directions

Emerging research combines the following aspects:

- **Meta-learning**: Combining MTL and continual learning with meta-learning strategies, which involve learning to learn, can potentially lead to systems that rapidly adapt to new tasks with minimal data and without forgetting
- **Neurosymbolic AI**: Integrating these models with neurosymbolic AI, which combines neural networks with symbolic reasoning, offers a pathway to more robust understanding and reasoning capabilities, further bridging the gap between AI and human-like intelligence

The ongoing research and development in multitasking and continual learning models is paving the way for AI systems that are not only more efficient and capable across a range of tasks but also adaptable and resilient in the face of new challenges. This progress underscores the continuous push toward AI systems that can seamlessly integrate into dynamic real-world environments, offering solutions that are both innovative and practically viable.

Integration of multitasking and continual learning

Integrating multitasking and continual learning could lead to models that are not only capable of learning multiple tasks simultaneously but also capable of adapting to new tasks over time without forgetting previous knowledge. This integration represents a significant step toward more flexible, efficient, and human-like AI systems.

Research continues to explore ways to improve these models, including developing more effective strategies for knowledge transfer, preventing catastrophic forgetting, and efficiently managing computational resources. The synergy between multitasking and continual learning models is poised to drive advancements in AI, enabling the development of more robust, adaptable, and intelligent systems.

Case study – implementing multitasking and continual learning models for e-commerce personalization

Let's consider a case study that focuses on the retail industry.

Background

In a hypothetical case study, an e-commerce giant aimed to refine its customer experience by providing personalized shopping experiences. The objective was to develop a system capable of handling various aspects of customer interaction, from product recommendations to customer service inquiries.

Challenge

The challenge was twofold: the model needed to manage multiple tasks relevant to the e-commerce environment and also adapt to evolving customer behaviors and inventory changes over time without forgetting previous interactions.

Solution – multitasking and continual learning models

To tackle this, the company adopted a combination of multitasking and continual learning models. This involved the following phases:

- **Phase 1 – multitasking model development**:

  - **Integrated task learning**: The model was designed to handle product recommendations, sentiment analysis from customer reviews, and customer service inquiries concurrently

  - **Shared learning architecture**: Early neural network layers were trained to recognize common patterns in customer data, while later layers were dedicated to task-specific processing

- **Phase 2 – implementing continual learning**:

  - **Dynamic data incorporation**: The system was equipped to continuously incorporate new customer interaction data, learning from recent trends and preferences

  - **Replay mechanisms**: To prevent catastrophic forgetting, the model periodically revisited previous customer data to retain historical knowledge

  - **Regular model updates**: The model architecture allowed for periodic updates without complete retraining, adapting to new products and customer service scenarios

Results

The implementation led to a robust, multifaceted AI system that did the following:

- Increased product recommendation accuracy by 40%, enhancing cross-selling and upselling opportunities

- Improved customer service response times by 30%, with more accurate and helpful responses

- Demonstrated significant retention of customer preferences over time, leading to a more personalized shopping experience

Impact

The multitasking and continual learning models had substantial impacts:

- **Customer experience**: The system's ability to provide accurate recommendations and timely customer service improved overall customer satisfaction

- **Business insights**: Continual learning from customer interactions provided valuable insights into buying patterns, helping inform inventory and marketing strategies

- **Operational efficiency**: By handling multiple tasks within one model, the company streamlined its operations, reducing the need for separate systems and teams

Conclusion

This case study highlights the successful application of multitasking and continual learning models in an e-commerce setting, addressing complex customer interaction needs while adapting to an ever-changing market landscape. The combination of these AI methodologies provided a competitive edge, not only by enhancing the user experience but also by offering a scalable solution for personalized customer engagement that evolves with consumer behavior and market demands.

Next, we'll introduce a case study that addresses training an LLM for a specialized domain.

Case study – training an LLM for a specialized domain

Training an LLM for a specialized domain involves a series of intricate steps, meticulous planning, and strategic implementation to ensure the model's effectiveness in understanding and generating text relevant to the specific field. This process can be dissected into several phases, each of which is crucial for the model's development and eventual performance. Let's explore a hypothetical case study that illustrates how an LLM could be trained for a specialized domain, such as medical research:

- **Phase 1 – defining the objectives and scope**:

  - **Objective setting**: The first step involves clearly defining the objectives of the LLM within the specialized domain. For instance, in the medical research domain, the model could aim to assist in generating medical research papers, interpreting clinical study results, or answering medical inquiries.

  - **Scope determination**: Deciding the scope involves specifying the breadth and depth of knowledge the model needs. For a medical research LLM, the scope could range from general medical knowledge to specific sub-fields, such as oncology or genomics.

- **Phase 2 – data collection and preparation**:

  - **Data sourcing**: Collecting a comprehensive and high-quality dataset is paramount. For medical research, this could involve gathering a wide array of texts, including research papers, clinical trial reports, medical journals, and textbooks.

  - **Data cleaning and preprocessing**: The collected data must be cleaned and preprocessed to remove irrelevant information, correct errors, and standardize formats. This step is crucial to ensure the model learns from accurate and relevant data.

  - **Data annotation**: Annotating data with metadata, such as topics or categories, can help in training more refined and context-aware models. For specialized domains, expert annotators are often required to ensure the accuracy of annotations.

- **Phase 3 – model selection and training**:

  - **Choosing a base model**: Selecting an appropriate base model or architecture is critical. For a specialized LLM, starting with a pre-trained model that has been trained on a broad dataset and then fine-tuning it for the specialized domain often yields the best results.

  - **Fine-tuning for the domain**: Fine-tuning involves adjusting the pre-trained model on the specialized dataset. This step adapts the model's weights and biases to better reflect the nuances of the domain's language and knowledge.

  - **Evaluation and iteration**: Continuously evaluating the model's performance through metrics such as accuracy, fluency, and relevance is essential. Feedback loops help in iteratively refining the model through additional training or data adjustments.

- **Phase 4 – implementation and deployment**:

  - **Integration into applications**: Deploying the trained LLM involves integrating it into applications or workflows, where it can assist professionals in the domain. For medical research, this could be within systems for drafting research papers, providing clinical decision support, or educational tools.

  - **Monitoring and updating**: Post-deployment, the model's performance must be monitored to ensure it continues to meet the required standards. Over time, incorporating new data and research findings into the training dataset helps the model remain current and valuable.

Challenges and considerations

The following are some issues that you should be aware of:

- **Ethical and privacy concerns**: In sensitive domains such as medicine, ethical considerations and patient privacy are paramount. Ensuring data de-identification and compliance with regulations such as HIPAA is essential.

- **Bias and fairness**: Training data in specialized domains can contain biases. Active measures must be taken to identify and mitigate these biases to ensure the model's outputs are fair and unbiased.

- **Domain expertise**: Involving domain experts throughout the training process is critical for ensuring the relevance and accuracy of the model's outputs. Their insights can guide data collection, annotation, and evaluation processes.

Conclusion

Training an LLM for a specialized domain is a complex, multidisciplinary endeavor that requires careful planning, domain expertise, and continuous iteration. The process from defining objectives to deploying and maintaining the model involves various challenges, including ethical considerations, data quality, and model bias. However, when executed well, the result can be a powerful tool that enhances decision-making, accelerates research, and improves outcomes within the specialized domain. The hypothetical case study of training an LLM for medical research highlights the potential of specialized LLMs to contribute significantly to their respective fields, underscoring the importance of targeted training and the nuanced approach required to harness the full capabilities of LLMs.

Summary

Transfer learning and fine-tuning have revolutionized ML and NLP by significantly improving the training process of models. These methodologies allow pre-trained models to be adapted to new tasks, substantially reducing the need for large amounts of labeled data and decreasing computational resources. This efficiency gain shortens training times and reduces the reliance on extensive datasets while enhancing model performance through higher accuracy and better generalization capabilities. Fine-tuning builds upon this by tailoring the pre-trained model to specific domains. However, it comes with the risk of overfitting, which can be managed through strategic tuning and rigorous validation. These approaches democratize AI technology, making advanced modeling accessible to a wider range of users and accelerating the pace of innovation in the field.

Complementing these, curriculum learning refines the training approach by increasing task complexity incrementally, which mirrors human learning patterns and bolsters both the efficiency of learning and the model's ability to generalize. Implementing these methods requires carefully selecting appropriate pre-trained models and preparing task-specific data meticulously, ensuring the new models are finely tuned to the new tasks' requirements. Despite challenges such as domain mismatch and the complex nuances of fine-tuning, the benefits these strategies offer outweigh such hurdles. Moreover, the integration of multitasking and continual learning models, which allow systems to handle multiple tasks and adapt over time, further enhances AI's capabilities. These models employ shared architectures for efficiency and dynamic strategies to prevent catastrophic forgetting, enabling continuous adaptation and learning. Together, they provide a robust foundation for future AI systems that are adaptable, efficient, and capable of lifelong learning, promising to further advance AI's role in a multitude of complex and diverse tasks.

In the next chapter, we'll dive deeper into fine-tuning LLMs for specific applications.

5

Fine-Tuning LLMs for Specific Applications

In this chapter, we'll focus on the versatility of LLMs and specify fine-tuning techniques tailored to a variety of NLP tasks. From the intricacies of conversational AI to the precision required for language translation and the subtleties of sentiment analysis, you'll learn how to customize LLMs for nuanced language comprehension and interaction, equipping them with the skills they need to meet specific application needs.

In this chapter, we're going to cover the following main topics:

- Incorporating LoRA and PEFT for efficient fine-tuning
- Understanding the needs of NLP applications
- Tailoring LLMs for chatbots and conversational agents
- Customizing LLMs for language translation
- Sentiment analysis and beyond – fine-tuning for nuanced understanding

By the end of this chapter, you should be able to understand how to augment the adaptability of LLMs for a variety of NLP tasks, with a clear focus on fine-tuning practices tailored to distinct objectives.

Incorporating LoRA and PEFT for efficient fine-tuning

In the domain of NLP, fine-tuning large pre-trained models on specific tasks or domains can be computationally expensive and time-consuming. The **Low-Rank Adaptation (LoRA)** and **Parameter-Efficient Fine-Tuning (PEFT)** techniques address these challenges by reducing the number of parameters that need to be updated during fine-tuning, thus making the process more efficient and accessible. Let's review them in detail.

LoRA

LoRA is a technique that fine-tunes LLMs by introducing low-rank decomposition into the training process. Instead of updating all model parameters, LoRA modifies only a small subset, which significantly reduces the computational overhead. This approach is particularly beneficial when working with large models where updating all parameters would be infeasible due to resource constraints.

Let's look at some of the primary applications of LoRA:

- **Model personalization**: LoRA efficiently personalizes large pre-trained models for specific tasks or domains by fine-tuning only a small subset of parameters, enabling adaptation to niche applications without extensive computational resources.

- **Resource-constrained environments**: LoRA allows large models to be fine-tuned in resource-limited settings, such as on-edge devices or with restricted **graphics processing unit** (**GPU**) availability, enabling the deployment of sophisticated models in more accessible or mobile environments without requiring high-end hardware.

- **Multilingual and multimodal applications**: LoRA efficiently fine-tunes models for multilingual or multimodal tasks by selectively adapting them, making it ideal for creating multilingual chatbots or models that integrate text with visual data.

- **Rapid prototyping**: LoRA offers researchers and developers a quick, resource-efficient way to experiment with model fine-tuning. This is especially valuable in settings that require multiple iterations to explore various hypotheses or model architectures.

- **Custom AI solutions for enterprises**: Enterprises can leverage LoRA to fine-tune large models for specific business needs, such as understanding industry-specific terminology or improving task performance, enabling customized solutions without requiring vast computational resources.

- **Low-resource language processing**: LoRA allows models to be adapted to low-resource languages by reducing computational and data requirements, making it easier to fine-tune models despite data scarcity.

- **On-device AI**: LoRA enables efficient fine-tuning of models so that they can run on devices with limited computational power, such as smartphones or IoT devices, enhancing AI capabilities on the device while improving user privacy and reducing reliance on cloud communication.

The following are the key benefits of LoRA:

- **Computational efficiency**: By updating fewer parameters, LoRA reduces the required computational resources, making it feasible to fine-tune large models on less powerful hardware

- **Faster training**: The reduced parameter set leads to quicker convergence during training, enabling faster iteration and deployment of fine-tuned models

- **Memory efficiency**: LoRA's low-rank matrices require less memory, allowing larger models to be fine-tuned on devices with limited memory capacity

PEFT

PEFT builds on the principles of LoRA but introduces additional methods to further enhance fine-tuning efficiency. PEFT techniques include adapter layers, prefix-tuning, and other strategies that focus on updating only a small portion of the model's parameters while keeping the majority of the pre-trained weights frozen.

A few applications of PEFT are as follows:

- **Domain-specific adaptation**: PEFT is ideal for scenarios where models need to be adapted to specific domains, such as legal or medical language, without requiring full retraining.
- **Resource-constrained environments**: PEFT allows for effective model adaptation in environments with limited computational resources, such as edge devices or mobile applications.
- **Leveraging human feedback**: Incorporating human feedback into the fine-tuning process is crucial for aligning LLMs with human values, preferences, and ethical standards. Two key techniques for integrating human feedback are **proximal policy optimization** (**PPO**) and **direct preference optimization** (**DPO**).

The following are the key benefits of PEFT:

- **Domain-specific adaptation**: PEFT is ideal for adapting models to specific domains, such as legal or medical language, without the need for full retraining, making it highly efficient for specialized applications.
- **Reduced computational costs**: By fine-tuning only a small subset of parameters, PEFT significantly lowers computational requirements, making it feasible to adapt large models even in resource-constrained environments.
- **Faster fine-tuning**: PEFT accelerates the fine-tuning process, allowing models to be adapted quicker to new tasks or datasets. This is particularly beneficial in dynamic environments where rapid deployment is essential.
- **Maintained model integrity**: Since PEFT keeps the majority of the pre-trained model's weights frozen, it preserves the general knowledge of the original model while efficiently adapting to new tasks, ensuring both versatility and specialization.
- **Scalability**: PEFT's approach to fine-tuning allows for scalable adaptation across various tasks and domains without the need for extensive computational infrastructure, making it suitable for diverse applications.

PPO

PPO is a reinforcement learning algorithm that's used to fine-tune LLMs by optimizing a policy that aligns model outputs with human feedback. In the context of LLMs, this feedback is often provided by human evaluators, who rank or score model outputs based on their quality, relevance, or ethical considerations.

Here's how PPO enhances LLM fine-tuning:

- **Policy refinement**: PPO refines the model's behavior by iteratively adjusting the policy so that it produces outputs that are more aligned with human preferences

- **Stability**: PPO maintains a balance between exploration and exploitation, ensuring that the model continues to improve while avoiding drastic changes that could degrade performance

- **Scalability**: PPO can be applied to large-scale LLMs, enabling continuous improvement as more feedback is collected

DPO

DPO is another approach to incorporating human feedback, but unlike PPO, it focuses on directly optimizing model parameters based on preference data. This technique involves training the model on pairs of outputs, where one is preferred over the other, and adjusting the model's parameters to increase the likelihood of generating the preferred output.

Here are the advantages of DPO:

- **Simplicity**: DPO is simpler to implement than reinforcement learning approaches such as PPO, making it easier to integrate into existing fine-tuning pipelines

- **Direct feedback utilization**: By directly using preference data, DPO provides a straightforward way to align model behavior with user expectations

- **Flexibility**: DPO can be applied to various tasks, from generating coherent text to ensuring ethical AI outputs, by directly reflecting human preferences in the model's fine-tuning process

Integrating LoRA, PEFT, PPO, and DPO into fine-tuning practices

Combining the efficiency of LoRA and PEFT with the alignment capabilities of PPO and DPO offers a powerful approach to fine-tuning LLMs. By reducing the computational burden and simultaneously incorporating human feedback, developers can create models that are both highly efficient and closely aligned with human values.

Here are a few practical considerations:

- **Task-specific fine-tuning**: Utilize LoRA and PEFT to efficiently adapt models to specific tasks while maintaining performance

- **Feedback-driven refinement**: Implement PPO and DPO to iteratively refine models based on human feedback, ensuring outputs are ethical, relevant, and user-friendly

- **Continuous learning**: Employ these techniques in a continuous learning framework, where models are regularly updated and improved based on new data and feedback

With these techniques in mind, it's essential to consider the specific needs of NLP applications.

Understanding the needs of NLP applications

NLP applications are designed to enable machines to understand and interpret human language in a valuable way. Let's look at some of the core needs that these applications typically aim to address.

Computational efficiency

Computational efficiency is a critical factor in the development and deployment of NLP applications due to the following reasons:

- **Large datasets**: NLP models are typically trained on vast amounts of data. Efficiently handling and processing these datasets is essential for training models within a reasonable timeframe and without prohibitive costs.

- **Complex models**: State-of-the-art NLP models, such as Transformers, involve millions or even billions of parameters. Managing such complexity requires substantial computational power and efficient algorithms.

- **Real-time processing**: Many NLP applications, such as virtual assistants, translation services, and chatbots, need to process language data in real time. Computational efficiency is crucial for meeting the latency requirements for a good user experience.

- **Energy consumption**: The energy required to train and run large NLP models has financial and environmental impacts. Efficient use of computational resources can help mitigate these concerns.

- **Scalability**: NLP applications often need to scale to accommodate a growing number of users or an increasing volume of data. Efficient computational practices enable this scalability without linear increases in cost or resources.

- **Cost**: Computational resources are expensive. Optimizing the efficiency of these resources can significantly reduce the costs associated with training and deploying NLP models.

- **Software libraries and frameworks**: Using optimized libraries and frameworks such as TensorFlow, PyTorch, and Hugging Face Transformers can boost computational efficiency. These tools are designed for performance and integrate well with hardware accelerators, speeding up model training and inference.

- **Inference optimization**: Inference optimization techniques such as model compression and runtime adjustments enhance NLP model efficiency, reduce latency, and improve scalability, especially in real-time applications.

- **Streaming data**: Streaming data techniques let NLP models process continuous data efficiently by handling it in small increments, reducing latency. This is ideal for real-time applications such as live sentiment analysis or chatbots.

Several strategies can be employed to achieve computational efficiency in NLP applications:

- **Model optimization**: Techniques such as pruning, quantization, and knowledge distillation can reduce the size of NLP models without a significant loss in performance, leading to faster and less resource-intensive operations

- **Hardware accelerators**: Using specialized hardware such as GPUs, **tensor processing units** (**TPUs**), and **field-programmable gate arrays** (**FPGAs**) can speed up the training and inference processes

- **Efficient algorithms**: Implementing algorithms that can process data more quickly and with fewer computational steps can lead to more efficient NLP applications

- **Parallel processing**: Distributing the computation across multiple processors or machines can greatly reduce the time required for training and inference

- **Caching**: Storing frequently accessed data in fast-access memory locations can reduce the time spent retrieving data during model training and inference

- **Batch processing**: Grouping data into batches allows for more efficient processing by taking advantage of the parallel nature of modern CPUs and GPUs

- **Cloud computing**: Leveraging cloud resources can provide on-demand access to powerful computational infrastructure, optimizing cost and efficiency for varying workloads

- **Neural processing units** (**NPUs**): NPUs are specialized processors that speed up neural network execution, making NLP applications faster and more energy-efficient, especially on mobile or edge devices

- **Digital signal processors** (**DSPs**): DSPs, optimized for signal data such as audio and images, can also boost NLP tasks by handling feature extraction or text pre-processing, offloading work from the main processor and improving efficiency

- **Specialized AI accelerators (for example, Cerebras)**: Specialized AI accelerators from Cerebras Systems provide exceptional power for AI workloads, handling massive models with billions of parameters and cutting training time and energy use for large-scale NLP models

By focusing on computational efficiency, developers can build NLP applications that are not only powerful and accurate but also economical and environmentally sustainable. This is essential for the widespread adoption and long-term viability of NLP technology.

Domain adaptability

Domain adaptability in NLP applications refers to the capacity of these systems to understand and process language that's specific to particular fields or industries. This adaptability is crucial because language use, such as terminology, syntax, and semantics, can vary greatly from one domain to another. For instance, the way language is used in medical reports is vastly different from language in legal documents or everyday conversations.

Here are some key aspects of domain adaptability in NLP:

- **Specialized terminology**: Different fields have their own sets of jargon and technical terms that may not be used in general language or may have different meanings in a specialized context.

- **Unique linguistic structures**: Certain domains may use unique linguistic structures or syntax. For example, legal documents often contain long, complex sentences with a specific structure that can be quite different from other forms of writing.

- **Contextual meaning**: Words and phrases might have specific meanings within a domain that are not apparent to those outside it. NLP systems must be able to discern these domain-specific meanings.

- **Implicit knowledge**: Domains often have implicit knowledge that practitioners are familiar with but that may not be explicitly stated in text. NLP systems need to incorporate this background knowledge to fully understand domain-specific texts.

- **Regulatory compliance**: Some domains have regulatory requirements that dictate how information is processed and communicated. NLP applications must be adaptable to comply with these regulations.

- **Data scarcity**: High-quality, domain-specific datasets may be scarce or expensive to obtain, making it challenging to train NLP models that require large amounts of data.

- **Customization of model components**: Customizing NLP models by tailoring architectures, fine-tuning, and creating domain-specific embeddings enhances adaptability and accuracy in specialized fields. Regular updates and domain expertise integration keep the system relevant and effective.

To achieve domain adaptability, the following strategies are often employed:

- **Transfer learning**: Leveraging pre-trained NLP models on general data and then fine-tuning them on a smaller, domain-specific dataset

- **Custom datasets**: Creating or curating large datasets with domain-specific texts to train or fine-tune NLP models

- **Expert involvement**: Involving subject matter experts in the development process to ensure that the NLP system captures domain-specific knowledge accurately

- **Ontologies and knowledge bases**: Integrating structured domain knowledge through ontologies or knowledge bases can help NLP applications understand and generate domain-specific content

- **Continuous learning**: Implementing mechanisms for continuous learning from new domain-specific data as it becomes available, allowing the NLP system to stay up to date with evolving language use in the domain

- **Hybrid models**: Combining rule-based and machine learning approaches to handle both the predictable and variable aspects of domain-specific language

- **Custom tokenization and embeddings**: Customizing tokenization and developing domain-specific embeddings allow NLP models to capture unique linguistic features and improve understanding of domain-specific terms and their relationships

- **Model customization**: Adapting NLP model architecture to a specific domain by adjusting network depth, tweaking hyperparameters, or incorporating domain-specific features is crucial for achieving high performance and alignment with field complexities

- **Domain-specific augmentation**: Domain-specific data augmentation techniques, such as generating synthetic data that mimics real-world scenarios, expanding limited datasets, and improving the model's ability to generalize within the domain

Ensuring domain adaptability allows NLP applications to be used effectively across a wide range of specialized fields, such as healthcare, law, finance, and technical support, thus extending their utility and effectiveness.

Robustness to noise

Robustness to noise is a critical characteristic for NLP applications, allowing them to maintain high performance even when faced with irregular or unexpected data inputs. Let's take a closer look at this attribute.

Understanding noise in data

Noise in data refers to any kind of irregularity or anomaly that deviates from the standard or expected format. In the context of NLP, noise can come in various forms:

- **Typos**: Mistakenly altered characters within words that can change their meaning or make them unrecognizable to the system

- **Slang**: Informal language that may not be widely recognized or that may vary greatly between communities or over time

- **Grammatical errors**: Incorrect verb tenses, misplaced punctuation, wrong word order, or other mistakes that can confuse the intended meaning

- **Colloquialisms**: Everyday language that can include idioms or phrases particular to a specific region or group

- **Non-standard usage**: Creative or unconventional use of language, such as in poetry or certain types of advertising copy

- **Dialectal variations**: Differences in language use based on regional or cultural dialects

- **Speech disfluencies**: In spoken language applications, these can include hesitations, repetitions, and non-words such as "um" or "uh"

Strategies for building robust NLP systems

To build NLP systems that are robust to noise, developers can employ several strategies:

- **Data augmentation**: Artificially introducing noise into the training data can help the model learn to handle such irregularities

- **Preprocessing**: Implementing steps to clean and standardize data before it's fed into the model, such as spell-checking or expanding contractions

- **Contextual models**: Using models that take broader context into account can help disambiguate and correct errors based on surrounding text

- **Error-tolerant algorithms**: Algorithms designed to tolerate and even expect errors can maintain performance despite noisy inputs

- **Robust embeddings**: Word embeddings that group similar words close together in the vector space can help the model understand typos or slang as being close to their standard counterparts

- **Transfer learning**: Models pre-trained on large, diverse datasets often have inherent robustness to various kinds of noise due to their exposure to a wide range of language use

- **Regularization techniques**: Techniques such as dropout can prevent overfitting to the noise-free training data, enhancing the model's ability to generalize to noisy real-world data

- **Custom tokenization**: Designing tokenizers that can handle non-standard language use, such as splitting hashtags or understanding text-speak

- **Post-processing**: Implementing rules or additional models that can clean up or correct the outputs of the primary NLP model

- **User feedback**: Allowing systems to learn from user corrections and feedback to improve robustness over time

Benefits of noise robustness

NLP applications that can effectively manage noisy data are generally more user-friendly and accessible. They can be deployed in a wider range of real-world environments and are better at understanding and engaging with users in natural, informal settings. This resilience to noise is especially important in applications such as voice-activated assistants, automated customer service, content moderation, and social media analysis, where the inputs are highly varied and unpredictable.

So, robustness to noise is essential for the reliability and versatility of NLP systems, ensuring that they can perform well in the face of the messy, unstructured language data that is typical of human communication.

Scalability

Scalability in NLP applications refers to the capability to handle growing amounts of data and increasingly complex tasks efficiently, without a compromise in performance. As the use of NLP expands in various fields, from business intelligence to social media analytics, the ability to scale becomes a critical component of system design.

Benefits of scalability

Various benefits of scalability ensure efficient growth and adaptability to changing demands and market dynamics:

- **Cost-effectiveness**: Scalable NLP applications can grow with user demand without necessitating a complete overhaul, thus optimizing costs

- **Flexibility**: Scalable systems can quickly adapt to changing requirements, whether due to an increase in data, users, or complexity of tasks

- **User satisfaction**: Maintaining speed and accuracy despite growing demand ensures a consistent and satisfactory user experience

- **Market adaptability**: Scalable NLP applications can more readily adapt to market changes and accommodate new data sources and user needs

Challenges in scalability

Scalability poses several challenges for NLP systems:

- **Data volume**: As datasets grow in size, NLP systems must process and analyze data without significant slowdowns.

- **Concurrent users**: NLP services may face a large number of simultaneous users, thereby requiring concurrent processing without latency issues.

- **Model complexity**: More sophisticated NLP models tend to have more parameters, which can be computationally expensive and harder to scale.

- **Diverse data**: NLP applications must handle a variety of data types and languages, which can introduce complexity as they scale.

- **Distributed systems**: To tackle scalability challenges from large datasets and high user concurrency, NLP systems often use distributed environments for parallel task processing across multiple machines. This enhances throughput but introduces challenges in synchronization, fault tolerance, and data distribution.

- **Scalability of algorithms**: Ensuring NLP algorithms are scalable is crucial for maintaining performance as the system grows. It requires handling increasing data volumes and user requests efficiently, optimized for parallel execution and workload distribution across multiple processors or nodes.

Strategies for scalability

The following strategies can be implemented for scalability:

- **Efficient algorithms**: Optimizing algorithms for performance can reduce computational requirements, allowing for faster processing of larger datasets

- **Parallel processing**: Utilizing multithreading and distributed computing to perform parallel data processing can significantly improve scalability

- **Cloud computing**: Leveraging cloud resources can provide on-demand scalability, allowing systems to adapt to varying workloads with ease

- **Load balancing**: Distributing workload across servers can help manage the flow of data, ensuring stable performance as demand increases

- **Microservices architecture**: Building NLP applications as a collection of loosely coupled services can allow different components to scale independently as needed

- **Hardware acceleration**: Using specialized hardware such as GPUs can speed up computations, particularly for model training and inference tasks

- **Caching**: Storing frequently accessed data in cache memory can reduce the time taken to access this data, improving response times

- **Data sharding**: Segmenting large datasets into smaller, more manageable pieces can help maintain performance as the overall volume of data increases

- **Elastic resources**: Implementing systems that automatically adjust the amount of computational resources based on the current demand can ensure consistent performance

- **Optimized storage**: Efficient data storage solutions can speed up data retrieval times, which is crucial for large-scale NLP tasks

- **Batch processing**: Grouping data processing tasks into batches can optimize the use of computational resources

- **Monitoring and autoscaling**: Continuously monitoring system performance and automatically scaling resources can help maintain efficiency as user demand fluctuates

In summary, scalability is a vital characteristic of NLP systems that ensures they remain efficient and effective as they grow. By addressing the challenges of scalability with strategic planning and technological solutions, NLP applications can continue to deliver high-quality insights and services to an expanding user base.

Multilinguality

Multilinguality in NLP applications is a key feature that allows these technologies to operate across different languages, which is essential for global reach and accessibility. Let's take a detailed look into multilinguality in the context of NLP.

The significance of multilinguality

Multilinguality stands as a cornerstone in modern NLP systems that's vital for the following aspects in an increasingly connected society:

- **Global communication**: In an interconnected world, the ability to communicate and process information in multiple languages is crucial for individuals and businesses to reach a broader audience

- **Cultural inclusivity**: Multilingual NLP systems ensure that non-English speakers and those who speak minority languages are not left out, promoting inclusivity

- **Cross-cultural exchange**: These systems facilitate the exchange of information across cultural boundaries, fostering international collaboration and understanding

Benefits of multilingual NLP systems

Multilingual NLP systems confer numerous advantages, including the following for more comprehensive data analysis:

- **Broader reach**: Businesses and services can reach a global audience by providing support in multiple languages

- **Enhanced accessibility**: More people can access technology and information in their native languages, reducing language barriers

- **Improved user experience**: Users can interact with technology in the language they are most comfortable with, leading to better engagement and satisfaction

- **Diversity of input**: Multilingual systems can gather and understand a wider range of viewpoints and information, leading to more diverse and rich data analysis

Challenges in multilingual NLP

The following challenges are posed by multilinguality for NLP systems:

- **Language complexity**: Each language has its own set of grammatical rules, syntax, idioms, and nuances, making it challenging to create models that can accurately process multiple languages.

- **Resource availability**: While high-resource languages such as English have abundant data for training NLP models, low-resource languages may lack sufficient data, making it hard to develop robust models for them.

- **Contextual nuances**: Words and phrases can have different connotations and cultural references in different languages that NLP systems need to understand to maintain the meaning and sentiment of the text.

- **Script variations**: Different languages use different scripts, some of which, such as Chinese or Arabic, may require specialized processing due to their complexity or non-linearity.

- **Translation and alignment**: Translating content across multiple languages while preserving meaning, tone, and context is complex and proves challenging in aligning texts, especially between languages with different grammatical structures or word orders. In these cases, sophisticated alignment algorithms are required.

- **Interoperability and integration**: In multilingual environments, NLP systems must seamlessly integrate with various tools and platforms, overcoming challenges such as proprietary formats and diverse standards to ensure effective interaction and error-free communication.

Approaches to achieving multilinguality

Achieving proficiency in multiple languages is a multifaceted endeavor in the field of NLP that involves utilizing the following approaches, among others, to create systems capable of understanding and interacting across linguistic barriers:

- **Transfer learning**: Leveraging a model trained on one language to bootstrap performance on another, especially when the target language has limited training data

- **Cross-lingual embeddings**: Creating word or sentence embeddings that map semantically similar phrases across languages into proximate points in a high-dimensional space

- **Multilingual training**: Training NLP models on datasets that include multiple languages, which can help the model learn shared representations across languages

- **Language-specific tuning**: Fine-tuning a general multilingual model on language-specific data to improve performance for that particular language

- **Universal grammatical structures**: Utilizing knowledge of universal grammatical structures that apply across languages to inform model architecture and training

- **Zero-shot learning**: Developing models that can understand or translate languages they haven't been explicitly trained on by learning transferable knowledge from other languages

- **Multilingual data augmentation**: Augmenting training data with synthetic examples in multiple languages enhances multilingual NLP models by increasing diversity and coverage, especially for low-resource languages

- **Cultural and linguistic adaptation**: Incorporating cultural and linguistic nuances into NLP models ensures accurate translations that respect and reflect the cultural context, which is crucial for applications such as sentiment analysis

In conclusion, multilinguality is a fundamental aspect of modern NLP applications that aim to serve a global user base. Developing multilingual capabilities involves addressing linguistic diversity and complexity but yields significant benefits in terms of accessibility, inclusivity, and global reach. As NLP technology continues to advance, we can expect even more sophisticated multilingual systems that can navigate the subtleties of human languages more effectively.

User interaction

User interaction with NLP systems is a critical aspect that determines the usability and effectiveness of the technology. A well-designed **user interface** (**UI**) allows users to interact seamlessly with the underlying NLP capabilities, making complex technology accessible and functional for a broad audience.

Key components of user interaction in NLP

The key components of effective user interaction in NLP systems are as follows:

- **Intuitive design**: The interface should be designed so that it's intuitive to users of all levels of technical expertise. This involves clear and understandable instructions, feedback mechanisms, and a layout that's easy to navigate.

- **Responsive feedback**: Users should receive immediate and clear feedback from the system. For instance, when a user submits a query or command, they should know whether it's been understood and is being processed.

- **Error handling**: The system should gracefully handle errors, whether they're user input errors or system errors, and guide the user to the correct action without technical jargon that may confuse them.

- **Multimodal interaction**: For some applications, offering multimodal interfaces, including text, voice, and even gesture, can greatly enhance accessibility and ease of use.

- **Personalization**: NLP systems can improve user interaction by learning from individual user behavior and preferences to provide personalized experiences.

- **Consistency**: Ensuring that the NLP system has consistent behavior across different platforms and devices guarantees that users have a coherent experience, regardless of how they access the service.

- **Accessibility**: Interfaces should be designed with accessibility in mind so that users with disabilities can also interact with NLP applications. This includes considerations for screen readers, alternative input methods, and clear visual design.

- **Contextual awareness**: NLP systems should be context-aware, understanding the user's intent based on the interaction history and the current environment.

Challenges in designing for user interaction

Designing user interfaces for NLP systems presents distinct challenges, including the following:

- **Diverse user base**: Designing UIs that cater to users with different language skills, cultural backgrounds, and technological fluency can be challenging

- **Complex functions**: NLP capabilities can be highly complex and making them understandable and usable for the average user requires thoughtful UI/UX design

- **Feedback loops**: Creating effective feedback loops that help users understand the system's actions and improve their future interactions requires careful design and testing

- **User preferences**: Incorporating user preferences into NLP system design, such as language, tone, and interaction style, is crucial for creating personalized experiences and requires adaptable design frameworks

- **Learning over time**: Designing NLP systems that adapt to changing user behaviors and preferences over time adds complexity, requiring sophisticated algorithms and a design approach for continuous learning and refinement

Strategies for effective user interaction

Effective user interaction within NLP systems can be achieved through several key strategies:

- **User-centered design**: Engaging with potential users during the design process to understand their needs and preferences

- **Iterative design**: Continuously testing and refining the interface based on user feedback

- **Simplification**: Breaking down complex NLP tasks into simpler, user-friendly steps

- **Visualization**: Using graphical elements to represent data and results can make it easier for users to understand and interact with the system

- **Natural language feedback**: Using natural language to communicate with users can make interactions more comfortable and less formal

Impact of good user interaction

Good user interaction design in NLP systems is pivotal and includes the following aspects:

- **Increased adoption**: An easy-to-use interface can lead to wider adoption of the NLP application

- **Enhanced productivity**: Efficient user interaction can save time and reduce the learning curve, leading to increased productivity

- **User satisfaction**: Positive user experience can lead to higher satisfaction and retention rates

- **Cost reduction**: Well-designed user interactions can reduce the need for extensive user support and training, lowering operational costs

- **Efficiency gains**: Streamlined user interfaces contribute to faster task completion and more efficient use of system resources, enhancing overall efficiency

In conclusion, designing user interfaces for NLP systems is a crucial component that affects the overall user experience. By focusing on user-friendly design principles and considering the needs and behaviors of users, developers can create NLP applications that are not only powerful but also accessible and enjoyable to use.

Ethical considerations

Ethical considerations in the development and deployment of NLP applications are essential to ensure that these technologies are used responsibly and don't perpetuate or exacerbate social inequalities or biases. Let's review the main points related to ethical considerations in NLP.

Bias and fairness

Addressing bias and ensuring fairness in NLP is critical. Let's take a closer look:

- **Data bias**: NLP models can inadvertently learn and replicate biases present in their training data. For example, if a dataset contains gender biases, the model may produce outputs that are unfairly biased toward one gender.

- **Algorithmic fairness**: Ensuring that NLP algorithms treat all groups of people fairly is critical. This means that decisions, predictions, or recommendations made by these systems should not be unfairly discriminatory based on attributes such as race, gender, age, or sexual orientation.

- **Representation**: It's important to have diverse representation in datasets to avoid excluding minority voices and perspectives.

Transparency and accountability

In the realm of NLP, the imperatives of transparency and accountability are paramount, with an emphasis on the following:

- **Explainability**: There's a growing demand for NLP systems to be able to explain their decisions or outputs in understandable terms. This transparency is important for building trust and for users to be able to contest decisions they believe are incorrect.

- **Accountability**: When NLP applications are used in decision-making processes that affect people's lives, it's vital to establish clear lines of accountability. This includes being able to identify and correct errors when they occur.

Privacy

In NLP, safeguarding privacy is crucial, necessitating stringent data protection measures and robust anonymization methods to secure personal information in compliance with legal standards:

- **Data privacy**: NLP systems often process sensitive personal information. Ensuring that this data is handled securely and in compliance with privacy laws (such as GDPR) is critical.

- **Anonymization**: Techniques to anonymize data are important to prevent the inadvertent revelation of personal information when NLP technologies are applied to large datasets.

Consent and autonomy

In the domain of NLP, emphasizing consent and autonomy is fundamental and requires the following:

- **Informed consent**: Users should be informed about how their data will be used and must give their consent for its use, especially when personal data is involved

- **User control**: Users should have some degree of control over how their data is used and the ability to opt out of data collection processes

Social impact

Addressing the social impact of NLP technologies demands a commitment to the following, ensuring respectful communication and equitable access for all users:

- **Cultural sensitivity**: NLP systems should be designed with an awareness of cultural differences and the potential for miscommunication or offense

- **Accessibility**: Ensuring that NLP technologies are accessible to people with disabilities is also an ethical concern as these tools shouldn't create or reinforce barriers to information

Design and development

The design and development of NLP systems thrive on the following to ensure ethical considerations are integrated throughout the process:

- **Interdisciplinary approach**: Ethical NLP development benefits from the input of experts from various fields, including social science, law, and humanities, not just technology

- **Stakeholder engagement**: Engaging with stakeholders, including potential users and those affected by NLP applications, can provide insights into ethical concerns and how to address them

Regulations and standards

The following are relevant concerning regulations and standards:

- **Adherence to standards**: There are ethical standards and guidelines set by professional organizations and regulatory bodies that developers should adhere to

- **Monitoring and evaluation**: Continuously monitoring and evaluating NLP applications for ethical compliance is necessary, as is the willingness to make changes based on these evaluations

Addressing ethical considerations in NLP requires a proactive approach throughout the entire life cycle of the technology, from design to deployment and beyond. By considering these ethical issues, developers and organizations can help ensure that NLP technologies are used in ways that are fair, just, and beneficial to society.

Interoperability

Interoperability is a key aspect of NLP applications, allowing them to function seamlessly within a larger ecosystem of software and workflows. This section will provide a comprehensive overview of interoperability within the context of NLP.

Definition and importance

Interoperability refers to the ability of different systems and organizations to work together (interoperate). For NLP applications, this means the ability to exchange and make use of information across various software platforms, tools, and data infrastructures.

Benefits of interoperability

Interoperability brings multifaceted benefits, such as the following:

- **Flexibility**: Interoperable systems are more flexible and can be more easily adapted to changing requirements or integrated with new technologies

- **Efficiency**: Interoperability reduces the need for data re-entry or conversion, saving time and reducing the potential for errors

- **Collaboration**: It enables different organizations and systems to collaborate and share data, leading to better decision-making and innovation

- **Scalability**: Interoperable systems can more easily scale as they can be expanded with components from different vendors that work together

- **User satisfaction**: For end users, interoperability leads to smoother workflows and a more cohesive experience as they can use different tools and systems together with less friction

Challenges in achieving interoperability

Achieving interoperability in NLP poses multiple challenges, including the following:

- **Diverse data formats**: NLP systems must handle a range of data formats, from structured data such as JSON or XML to unstructured text in various languages and formats

- **Different application programming interfaces (APIs)**: Integration often involves working with different APIs, each with its own set of protocols and data exchange formats

- **Varying standards**: There may be different industry standards or protocols that need to be adhered to, which can vary by region, sector, or type of data

- **Legacy systems**: Older systems may not have been designed with modern interoperability standards in mind, making integration more complex

Strategies for ensuring interoperability

To ensure interoperability within NLP applications, various strategies can be implemented:

- **Standardization**: Adhering to industry standards for data formats and APIs can greatly facilitate interoperability

- **Use of common protocols**: Employing widely-used protocols such as REST for web services ensures that NLP applications can easily communicate with other systems

- **Middleware**: Middleware can act as a bridge between different systems and data formats, translating and routing data as needed

- **Data wrappers**: Implementing wrappers can convert data from one format into another, allowing for smooth integration between systems that use different data structures

- **Service-oriented architecture (SOA)**: Designing systems with an SOA can ensure that individual components can be accessed and used by other systems without them needing to share the same technology stack

- **Microservices**: This involves building NLP applications as a suite of small, modular services, each running its own process and communicating through lightweight mechanisms, typically an HTTP resource API

- **Open standards**: Developing and using open standards for data exchange and APIs enhances the ability of different systems to work together

- **Documentation**: Providing clear and comprehensive documentation for APIs and data formats is crucial for enabling other developers to create interoperable systems

- **Testing and validation**: Regularly testing NLP applications to ensure they work as expected with other systems is essential for maintaining interoperability

In summary, interoperability is a critical feature for NLP applications to ensure they can be integrated into various digital environments. It allows data and functionality to be exchanged seamlessly across different systems, enhancing the value and usability of NLP technologies.

By fine-tuning LLMs to cater to these needs, developers can create highly effective NLP applications tailored to specific tasks, industries, or user requirements. The key to success lies in careful preparation, clear task definition, and ongoing model refinement. The next section deals specifically with tailoring LLMs for the particular tasks of chatbots and conversational agents.

Tailoring LLMs for chatbots and conversational agents

Tailoring LLMs for chatbots and conversational agents is a process that involves customizing these models so that they better understand, respond to, and engage with users in conversational contexts. Let's take a closer look at how LLMs can be tailored for such applications.

Understanding the domain and intent

Understanding the domain and intent is a crucial aspect of tailoring LLMs for applications such as chatbots and conversational agents. Let's take a closer look.

Domain-specific knowledge

Domain-specific knowledge in LLMs necessitates a focused approach to learning, ensuring depth in the specific field and the ability to keep updated with new developments. This approach includes the following aspects:

- **Tailoring to the domain**: LLMs typically have a broad understanding of language from being trained on diverse datasets. However, chatbots often need to operate within a specific domain, such as finance, healthcare, or customer service. Tailoring an LLM to a specific domain involves training it on a corpus of domain-specific texts so that it can understand and use the specialized terminology and knowledge effectively.

- **Depth of knowledge**: Domain-specific tailoring also means ensuring that the LLM can answer deeper, more complex queries specific to the domain. For example, a medical chatbot should understand symptoms, diagnoses, and treatments, while a financial chatbot should understand various financial products and economic terms.

- **Continual learning**: Domains evolve, with new terminology and practices emerging. Therefore, domain-specific chatbots must be capable of continual learning to update their knowledge base.

Intent recognition

Intent recognition is essential in NLP for discerning the following:

- **Understanding user queries**: Intent recognition is the process of determining what users want to achieve with their queries. This could range from seeking information, making a booking, getting help with a problem, or a myriad of other intents. Accurately recognizing intent is crucial for providing correct and useful responses.

- **Training on intent datasets**: Fine-tuning an LLM for intent recognition typically involves training on datasets that include a wide variety of user queries labeled with their corresponding intents. This training helps the model to learn the patterns in how users phrase different types of requests.

- **Handling ambiguity**: User queries can often be ambiguous and may be interpreted in multiple ways. LLMs must be trained to identify the most likely intent based on the context or ask clarifying questions when necessary.

- **Multi-intent recognition**: Sometimes, user queries may contain multiple intents. For instance, a user might ask a travel chatbot about weather conditions and car rentals in a single message. Fine-tuning for multi-intent recognition allows the chatbot to address each part of the query.

Integration with backend systems

For many applications, understanding the domain and intent is just the first step. The chatbot often needs to take action based on this understanding, such as retrieving information from a database or executing a transaction. This requires seamless integration with backend systems, which must be accounted for in the design and training of the chatbot.

Ethical and practical considerations

When fine-tuning LLMs, it's also important to consider ethical implications. This includes ensuring that the chatbot doesn't reinforce stereotypes or biases and respects user privacy.

In summary, fine-tuning LLMs for domain-specific knowledge and intent recognition is a multifaceted process that requires carefully considering the specific requirements of the domain, the nuances of user queries, and the need for ongoing learning and integration with other systems. This process ensures that chatbots and conversational agents can provide high-quality, relevant, and contextually appropriate interactions.

Personalization and context management

In enhancing user experience, personalization and context management are pivotal, with conversational agents designed to retain dialog context and LLMs customized for individual user engagement through learning and personalization. Let's take a closer look:

- **Maintaining context**: Conversational agents must maintain the context of a conversation over multiple exchanges, which requires memory and reference capabilities. LLMs can be tailored to remember previous parts of the conversation and reference this context in their responses.

- **Personalization**: To make interactions more engaging, LLMs can be customized to learn from previous interactions with users and to personalize the conversation based on the user's preferences and history.

Natural language generation

Natural language generation (**NLG**) is a critical aspect of LLMs that enables them to generate text that's coherent, contextually relevant, and similar to human language. When applied to chatbots and conversational agents, NLG plays a significant role in how these systems communicate with users. Let's take a detailed look at the key components.

Generating human-like responses

In crafting responses that emulate human dialog, LLMs undergo training on the following:

- **Conversational data training**: To produce responses that closely mimic human conversation, LLMs are trained on large datasets of real dialogs. This training helps the model understand a variety of conversational patterns, idioms, and the flow of natural discourse.

- **Understanding pragmatics**: Beyond the words themselves, human-like responses also require an understanding of pragmatics – the study of how context contributes to meaning. For instance, when a user says, "It's a bit chilly in here," a well-tuned chatbot might respond by suggesting how to adjust the temperature, recognizing the implicit request.

- **Techniques for naturalness**: Techniques such as reinforcement learning can be used to fine-tune the LLM's ability to generate responses that not only answer the user's query but also engage in a manner that's contextually and emotionally appropriate.

Variability in responses

In striving to enhance user engagement, LLMs employ strategies to do the following:

- **Avoid repetition**: Chatbots that always respond in the same way can quickly feel mechanical. By introducing variability in the responses, an LLM can make each interaction feel unique and more engaging.

- **Provide diverse responses**: This can be achieved through techniques such as beam search during the generation process, where the model considers multiple possible responses and selects one that's appropriate but perhaps less obvious or more varied.

- **Generate dynamic content**: LLMs can be designed to reference external and dynamic content sources, ensuring that responses are not only varied but also up-to-date and relevant to current events or user-specific data.

The importance of NLG in user experience

In crafting compelling user experiences, NLG plays a pivotal role by providing the following:

- **User engagement**: Human-like and varied responses can significantly improve user engagement as interactions with the chatbot become more enjoyable and less predictable

- **User trust**: When a chatbot can provide responses that seem thoughtful and well-considered, it builds trust with the user, who may feel more confident relying on the chatbot for information or assistance

- **Personalization**: NLG can be combined with user data to create personalized experiences, where the chatbot refers to past interactions or user preferences, further enhancing the natural feel of the conversation

Challenges and considerations

The following are some challenges and considerations regarding NLG:

- **Balance between consistency and variety**: While variability is important, it's also crucial to maintain consistency in the chatbot's tone and personality, which requires carefully calibrating the NLG process

- **Context retention**: In a long conversation, the chatbot must retain the context and ensure that variability in responses does not lead to loss of coherence or relevance

- **Cultural sensitivity**: Responses must be culturally sensitive and appropriate, which can be challenging when generating varied content for a global audience

In summary, the goal of fine-tuning NLG in LLMs for chatbots and conversational agents is to create systems that provide responses that are not only correct but also contextually rich, engaging, and reflective of human conversational norms. Achieving this level of sophistication in NLG contributes significantly to the overall user experience and effectiveness of conversational AI.

Performance optimization

Efficient performance optimization is vital for chatbots as it ensures the following:

- **Response latency**: For a smooth conversation, chatbots need to respond quickly. LLMs must be optimized for performance to minimize latency.

- **Resource efficiency**: Chatbots may be required to handle multiple conversations simultaneously, which demands that the underlying LLMs be resource-efficient.

Ethical and privacy considerations

In terms of ethical and privacy considerations, tailoring LLMs involves doing the following:

- **Avoiding harmful outputs**: Tailoring LLMs includes implementing safeguards against generating harmful, biased, or inappropriate content.

- **Privacy protection**: Conversational agents often deal with personal user data. LLMs should be tailored to respect user privacy and handle sensitive data according to privacy standards and regulations.

Continuous improvement

Continuous improvement in conversational agents involves implementing the following:

- **Feedback loops**: Implementing feedback mechanisms allows the LLM to learn from user interactions and continuously improve its conversational abilities

- **Monitoring and updating**: Regularly monitoring chatbot performance and updating the underlying LLM to reflect new data, trends, or feedback help maintain the relevance and effectiveness of conversational agents

By carefully tailoring LLMs to meet these requirements, developers can create chatbots and conversational agents that are more helpful, engaging, and enjoyable for users. The tailoring process involves not only making technical adjustments but also considering the ethical implications of deploying AI in user-facing applications.

The next section deals with tailoring LLMs for a different purpose – language translation.

Customizing LLMs for language translation

Customizing LLMs for language translation involves adapting and refining NLP systems to accurately translate text or speech from one language into another. This customization is essential for developing effective machine translation tools that can handle the nuances and complexities of different languages. Let's take an in-depth look at the process.

Data preparation

Data preparation for language translation involves the following aspects:

- **Parallel corpora**: A crucial step is gathering parallel corpora, which consist of large sets of text in two languages that are direct translations of each other. These corpora are used to train the model so that it understands how concepts and phrases in one language translate into another.

- **Domain-specific data**: For specialized translation tasks, such as legal or medical translations, it's important to include domain-specific vocabulary and phrases in the training data.

Model training

Model training for language translation often involves the following:

- **Neural machine translation (NMT)**: Modern translation models typically use neural networks, particularly sequence-to-sequence architectures, that can learn complex mappings from source to target languages

- **Transfer learning**: Leveraging pre-trained models on high-resource languages and then fine-tuning them on specific language pairs, especially if one of the languages has less data available

Handling linguistic nuances

Translating effectively requires the following:

- **Contextual understanding**: Translation models must grasp context to correctly translate homonyms and words with multiple meanings.

- **Grammar and syntax**: Different languages have different grammatical structures. The model must be able to reconstruct the correct syntax in the target language.

Quality and consistency

Quality and consistency are assessed and ensured with the following:

- **Evaluation metrics**: Using BLEU, METEOR, and other evaluation metrics to measure the quality of translations and guide model improvements

- **Post-editing**: Incorporating a human-in-the-loop for post-editing can improve translation quality, especially for nuanced or high-stakes content

Dealing with limitations

Addressing rare words and dialectical variations, such as through subword tokenization and cultural sensitivity, is crucial for overcoming translation limitations:

- **Rare words**: Customizing the model to handle rare words or phrases, possibly through subword tokenization strategies such as BPE

- **Language variants**: Accounting for dialects and language variants to ensure translations are accurate and culturally appropriate

Ethical and practical considerations

The following are some essential considerations to think about:

- **Bias mitigation**: Ensuring the model doesn't perpetuate or amplify biases present in training data

- **Confidentiality**: In scenarios where sensitive information is translated, maintaining confidentiality is critical

Continuous improvement

Continuous improvement in translation models is facilitated through the following:

- **Active learning**: The model can continue to improve by learning from corrections and feedback in an ongoing manner

- **Real-time learning**: Some systems are designed to learn from user interactions in real time, adapting to new phrases and usage patterns

By customizing translation models to address these aspects, developers can create sophisticated tools capable of translating text and speech with a high degree of accuracy and fluency. The goal is to produce translations that are not only grammatically correct but also contextually and culturally relevant.

Sentiment analysis and beyond – fine-tuning for nuanced understanding

Fine-tuning LLMs for sentiment analysis is an intricate process that aims to enhance the model's ability to detect and interpret the nuances of human emotion in text. Let's take a closer look at this process.

The basics of sentiment analysis

Sentiment analysis entails the following:

- **Polarity detection**: At its core, sentiment analysis involves determining the polarity of a piece of text, classifying it as positive, negative, or neutral

- **Emotion detection**: Beyond polarity, sentiment analysis can also involve detecting specific emotions, such as happiness, anger, or sadness

Challenges in sentiment analysis

Sentiment analysis faces challenges such as the following:

- **Contextual nuances**: The same word or phrase can convey different sentiments in different contexts. Fine-tuning LLMs to understand these nuances is crucial.

- **Sarcasm and irony**: Detecting sarcasm and irony requires a deep understanding of language and context as they often mean the opposite of the literal words used.

- **Cultural variations**: Sentiment expression can vary significantly across cultures, so models must be fine-tuned to understand these variations appropriately.

- **Generalization across text types**: Sentiment analysis models must generalize across diverse text types, adapting to different styles, lengths, and structures while maintaining accuracy in sentiment detection.

Fine-tuning for nuanced understanding

Fine-tuning for nuanced understanding involves the following:

- **Advanced training techniques**: Leveraging techniques such as transfer learning, where a model pre-trained on large datasets is further trained (fine-tuned) on sentiment-specific data

- **Domain-specific data**: Using domain-specific training data can help the model understand sentiments unique to certain fields, such as the financial or healthcare industry

- **Incorporating external knowledge**: Infusing the LLM with external knowledge sources, such as sentiment lexicons or encyclopedic databases, can improve its understanding of nuanced sentiment

Evaluation and adjustment

Assessing and refining sentiment analysis involves iterative feedback and the use of evaluation metrics such as accuracy and precision. This process is crucial for practical applications such as understanding customer feedback, market analysis, and product evaluation. Let's take a closer look:

- **Iterative feedback**: Using human-in-the-loop feedback to continuously refine the model's predictions

- **Evaluation metrics**: Employing metrics such as accuracy, precision, recall, and F1 score to evaluate the performance of the sentiment analysis and make necessary adjustments

Practical applications

The following are some of the practical applications of sentiment analysis:

- **Customer feedback**: Fine-tuned sentiment analysis models can help businesses understand customer sentiment from reviews, surveys, and social media posts

- **Market analysis:** In the financial sector, sentiment analysis can be used to gauge market sentiment and predict stock movements

- **Product analysis:** Companies can use sentiment analysis to monitor public sentiment about their products and services, identifying areas for improvement

- **Confusion matrix:** A confusion matrix can be used to evaluate the performance of sentiment analysis models by showing the true positives, false positives, true negatives, and false negatives

- **Receiver operating characteristic (ROC):** The ROC curve is a graphical representation that helps assess the trade-off between true positive and false positive rates in sentiment analysis models

- **Area under the curve (AUC):** The AUC score, which is derived from the ROC curve, provides a single metric to evaluate the overall performance of a sentiment analysis model, with higher values indicating better discrimination between positive and negative sentiments

Ethical considerations

Addressing ethical concerns in sentiment analysis involves the following aspects:

- **Bias mitigation**: Ensuring the model does not inherit or perpetuate biases from the training data, leading to skewed sentiment analysis

- **Privacy concerns**: Respecting user privacy when analyzing sentiment from personal communication or social media posts

Beyond sentiment analysis

The following are some advancements that go beyond traditional sentiment analysis:

- **Aspect-based sentiment analysis**: Breaking down sentiment to specific aspects of a product or service, such as the battery life of a phone or the comfort of a car
- **Emotion AI**: Developing models that can recognize a broader range of human emotions for applications in areas such as mental health support

In summary, fine-tuning LLMs for sentiment analysis requires a combination of advanced NLP techniques, comprehensive training data, iterative refinement, and a strong grasp of the ethical implications. The goal is to create models that can not only understand surface-level sentiment but also the deeper emotional undercurrents and subtleties present in human language.

Summary

In the landscape of NLP, computational efficiency and domain adaptability are paramount. NLP systems hinge on processing large datasets and complex models with efficiency, ensuring real-time interaction capabilities, and managing costs and energy consumption effectively. The scalability of these systems is crucial in handling the increasing data and user demand that can be achieved through model optimization, hardware accelerators, efficient algorithms, and cloud computing strategies. Such scalable systems provide the flexibility and user satisfaction necessary for widespread adoption, allowing them to adapt to market and data growth seamlessly.

Moreover, the ability to adapt to specific domains enriches the utility of NLP applications, allowing them to comprehend and process industry-specific language nuances. This includes mastering specialized terminology, recognizing unique linguistic structures, and understanding contextual meanings inherent to different fields. Achieving this level of adaptability often involves techniques such as transfer learning, the creation of custom datasets, and continuous learning mechanisms to keep pace with evolving domain-specific language use.

Sentiment analysis exemplifies the need for fine-tuning NLP models to capture the subtleties of human emotion in text. This fine-tuning is not just about detecting the polarity of sentiments but also the various shades of emotional expression. It involves advanced training techniques, domain-specific data training, and incorporating external knowledge sources for a nuanced understanding of sentiments. Ethical considerations such as bias mitigation and privacy protection are integral throughout this process, ensuring fairness and trustworthiness.

In conclusion, the development of NLP systems is a meticulous balancing act that requires paying attention to computational demands and the subtleties of human language. By addressing these core needs with sophisticated, adaptive, and ethical approaches, NLP applications are positioned to revolutionize how machines understand and interact with human language, making them indispensable tools across a myriad of applications.

In the next chapter, we'll move on and discuss LLM testing and evaluation.

6
Testing and Evaluating LLMs

After development, the next crucial phase is testing and evaluating LLMs, an aspect we'll explore in this chapter. We'll not only cover the quantitative metrics that gauge performance but also stress the qualitative aspects, including **human-in-the-loop** (**HITL**) evaluation methods. We'll also detail protocols while emphasizing the necessity of ethical considerations and the methodologies for bias detection and mitigation, ensuring that LLMs are both effective and equitable.

In this chapter, we're going to cover the following main topics:

- Metrics for measuring LLM performance
- Setting up rigorous testing protocols
- Human-in-the-loop – incorporating human judgment in evaluation
- Ethical considerations and bias migration

By the end of this chapter, you should have a comprehensive understanding of the crucial phase of testing and evaluating LLMs.

Metrics for measuring LLM performance

Metrics are essential for evaluating the performance of LLMs because they provide objective and subjective means to assess how well a model is performing relative to the tasks it's designed to complete. The following subsections present an expanded explanation of both quantitative and qualitative metrics used for LLMs.

Quantitative metrics

Quantitative metrics play a vital role in the evaluation of LLMs by providing objective, measurable indicators of performance. Let's review those metrics:

- **Perplexity**: Perplexity is a key metric in language modeling:

  - **Definition**: Perplexity is a measure of a model's uncertainty in predicting the next token in a sequence. It's a widely used metric in language modeling.

- **Calculation**: Perplexity is calculated as the exponentiated average negative log-likelihood of a sequence of words. A model that assigns higher probabilities to the actual words that appear next in the text will have lower perplexity.

- **Interpretation**: Lower perplexity indicates that the model is better at predicting the next word, suggesting a better understanding of the language structure.

- **Bilingual Evaluation Understudy (BLEU) score**): The BLEU score is a widely used metric for assessing the quality of machine-translated text:

  - **Definition**: BLEU is a metric for evaluating machine-translated text against one or more reference translations. It's one of the most common metrics for assessing the quality of text that has been machine-translated.

  - **Calculation**: The BLEU score evaluates the quality of text by comparing the n-grams of the machine-generated text to the n-grams of the reference text and counting the number of matches. These counts are then weighted and combined into a single score.

  - **Adjustments**: BLEU includes a brevity penalty to discourage overly short translations that might artificially inflate the score by having a high n-gram overlap.

- **Recall-Oriented Understudy for Gisting Evaluation (ROUGE)**: ROUGE also encompasses a set of metrics:

  - **Definition**: A collection of evaluation metrics, ROUGE is specifically designed for assessing machine translation and automatic summarization systems. It functions by contrasting generated translations or summaries with a set of benchmark summaries.

  - **Variants**: There are several variants of ROUGE, such as ROUGE-N (which compares n-grams), ROUGE-L (which uses the longest common subsequence), and ROUGE-S (which considers skip-bigrams, which are pairs of words in their sentence order, allowing for arbitrary gaps).

  - **Focus**: ROUGE can focus on recall, precision, or a balance of both (F-measure), depending on the variant used.

- **Accuracy**:

  - **Definition**: Accuracy is the fraction of predictions that a model gets right, including both true positives and true negatives, out of all the predictions made.

  - **Limitations**: In situations where classes are imbalanced, accuracy can be misleading. For example, in a dataset where 90% of the data belongs to one class, a model that always predicts that class will have high accuracy but poor predictive performance.

- **F1 score**:

  - **Definition**: The F1 score is a measure of a model's accuracy that considers both precision and recall. It's particularly useful when the class distribution is uneven.

- **Calculation**: The F1 score is the harmonic mean of precision (the accuracy of positive predictions) and recall (the ability of the classifier to find all positive instances).

- **Usefulness**: The F1 score is best used in scenarios where it's important to strike a balance between precision and recall, and when there's an uneven class distribution.

Using these metrics allows developers and researchers to quantify aspects of an LLM's performance and compare it with other models or benchmarks. While highly useful, these metrics should be complemented with qualitative evaluations to ensure a holistic understanding of the model's capabilities.

Qualitative metrics

Qualitative metrics are essential in evaluating the performance of LLMs because they provide a nuanced understanding of the model's outputs from a human perspective. These metrics go beyond the raw statistical measures to assess the quality and usability of the text generated by LLMs. Let's take a closer look at each of these qualitative metrics.

- **Coherence**:

  - **Description**: Coherence measures the logical flow of text and how each part connects to form a meaningful whole. It evaluates the text's structure and the clarity of transitions between sentences and paragraphs.

  - **Evaluation methods**: Human evaluators can rate coherence on a scale or through binary (yes/no) judgments. Automated methods might use discourse-level analysis to predict coherence, although these are less common and often less reliable than human evaluation.

- **Grammatical correctness**:

  - **Description**: This metric assesses the adherence of generated text to the rules of grammar. It includes syntax, punctuation, and morphological correctness.

  - **Evaluation tools**: Automated grammar checkers can identify many grammatical issues, but they may not catch more subtle errors or stylistic choices that affect readability. Hence, expert human evaluators are often used for a more accurate assessment.

- **Relevance**:

  - **Description**: Relevance is a measure of how well the text pertains to a given context, question, or topic. It's especially important in interactive applications such as conversational agents or question-answering systems.

  - **Assessment**: Human evaluators determine relevance by comparing the generated text against the context or prompt. They may consider whether the text is on-topic, answers the question posed, or addresses the user's intent appropriately.

- **Readability**:

  - **Description**: Readability indicates how easily a reader can understand the generated text. It encompasses factors such as sentence length, word difficulty, and the complexity of ideas presented.

  - **Assessment tools**: There are standardized tests for readability, such as the Flesch-Kincaid Grade Level or the Gunning Fog Index, which calculate scores based on sentence length and word complexity. Human evaluators can also provide subjective assessments of readability, especially for nuanced or complex texts.

Qualitative metrics demand a structured approach to ensure consistency and neutrality that involves detailed guidelines and trained evaluators. Though resource-intensive, they're crucial for evaluating LLMs based on user experience and practicality, aspects that quantitative metrics might miss. These metrics highlight a model's real-world efficacy beyond mere statistical performance.

Quantitative metrics, which are essential for initial model comparisons, offer automated, uniform performance indicators but may overlook language nuances. Qualitative evaluations, often through human judgment, fill this gap by assessing how human-like the model's output is.

Combining both types of metrics provides a comprehensive assessment of an LLM, covering both its statistical accuracy and the quality of its output as perceived by humans.

Setting up rigorous testing protocols

Setting up rigorous testing protocols is crucial for evaluating the effectiveness and reliability of LLMs. These protocols are designed to thoroughly assess the model's performance and ensure it meets the required standards before deployment. The following sections will provide a detailed exploration of how to set up such protocols.

Defining test cases

Defining test cases is a systematic approach to verifying that an LLM behaves as expected. Let's take a closer look at what goes into this process:

- **Typical cases**: These are scenarios that the model is expected to encounter frequently. For an LLM, typical cases might involve common phrases or questions that it should be able to understand and respond to accurately. The purpose is to confirm that the model performs well under normal operating conditions.

- **Boundary cases**: These are situations that lie at the edge of the model's operational parameters. For LLMs, boundary cases could involve longer-than-usual inputs, complex sentence structures, or ambiguities in language that are challenging but still within the scope of the model's capabilities. Testing boundary cases ensures that the model can handle inputs at the limits of what it was trained for.

- **Edge cases**: Edge cases are inputs that are unusual or rare, and they often reveal the model's behavior in exceptional situations. These might include slang, idiomatic expressions, or text with mixed languages. For LLMs, edge cases help us understand how the model deals with unexpected or unconventional inputs.

- **Negative cases**: These are tests where the model should ideally not take a certain action or make a specific prediction. For example, an LLM should not generate offensive content even if certain keywords are present in the input.

- **Performance cases**: It's also important to test how the model performs under different computational stress scenarios, such as processing a large volume of requests simultaneously or handling very large input texts.

When defining test cases for LLMs, consider the following aspects:

- **Diversity of data**: Include a variety of data sources, languages, dialects, and writing styles to ensure comprehensive coverage.

- **Relevance to use case**: Test cases should be relevant to the practical applications the LLM will be used for.

- **Automated and manual testing**: While many test cases can be automated, some will require manual assessment, especially when evaluating the nuances of language generation.

- **Iterative process**: As the model evolves, so should the test cases. They should be regularly reviewed and updated so that they match the model's expanding capabilities.

- **Documenting scenarios**: Maintain clear documentation for each test case, detailing the input, the expected output, and the rationale for the test.

- **Scalability**: Test cases should be scalable, allowing for automated testing as the number and complexity of cases grow.

In essence, defining test cases is a crucial step in validating that an LLM is robust, accurate, and ready for deployment, ensuring that it has been thoroughly evaluated across a spectrum of possible scenarios.

Benchmarking

Benchmarking is the process of setting performance standards that an LLM should meet or exceed. It involves comparing the model's performance against established baselines or standards. Here's an in-depth look at the benchmarking process:

- **Historical data**: Using historical performance data from previous versions of the model or similar models can provide insight into expected performance levels. For example, if an earlier version of an LLM achieved a certain BLEU score on a machine translation task, that score can serve as a benchmark for future versions.

- **Industry standards**: There are often well-established benchmarks within the AI and NLP communities for various tasks. For instance, standard datasets such as GLUE for natural language understanding or SQuAD for question-answering come with leaderboards that show the performance of top models. New models can be benchmarked against the leading scores on these leaderboards.

- **Custom benchmarks**: For specialized applications, you might need to create custom benchmarks that reflect the unique requirements of the task. For example, in a domain-specific language model, custom benchmarks could be based on the accuracy of the generated text, as assessed by domain experts.

- **Performance targets**: Benchmarks can also be set as specific performance targets. These targets might be derived from user requirements, business objectives, or technical constraints. For instance, a model might be required to generate responses within a certain timeframe to ensure user engagement.

- **Relative benchmarking**: Sometimes, it's useful to compare models relative to one another rather than against an absolute standard. This can be particularly helpful during development when iterating on different model architectures or training techniques.

- **Regression benchmarking**: In this context, regression doesn't refer to statistical regression but rather to software regression, where new changes might degrade performance. Regression benchmarks ensure that updates or improvements to the model do not cause a decline in performance on tasks that the model previously performed well on.

- **Extensibility**: Ensure that the benchmarks can be extended or adjusted as the capabilities of models and the tasks they are applied to evolve.

- **Reproducibility**: Benchmarks should be reproducible, meaning that they can be achieved consistently under the same testing conditions. This is crucial for the validity of the benchmarking process.

- **Documenting benchmarks**: Keep thorough records of the benchmarks used, including the source of the benchmark data, the rationale behind the benchmarks, and the methods used to measure against them.

Benchmarking is a continuous process that should accompany the life cycle of the model. It helps in goal setting, guides the development process, and ensures that the model meets the necessary standards before being deployed in a production environment.

Automated test suites

Automated test suites are a collection of tests that are executed by software to verify that different parts of a system, such as an LLM, are functioning correctly. These tests are designed to run automatically, without human intervention, and are a critical component of a robust testing strategy. Let's take a closer look at their importance and implementation:

- **Efficiency**: Automation allows a large number of tests to be executed in a short amount of time. This is particularly important for LLMs, which can be complex and require extensive testing to cover all functionalities.

- **Consistency**: Automated tests can be run repeatedly with the same conditions, ensuring that the results are consistent and reliable. This repeatability is vital for detecting when and how bugs are introduced.

- **Comprehensiveness**: Automated test suites can cover a wide range of test cases, including edge cases that might be overlooked during manual testing.

- **Integration testing**: Automated suites are not just for unit tests (which test individual components in isolation); they can also be used for integration tests, which verify how different parts of the model work together.

- **Regression testing**: They are ideal for regression testing, ensuring that new code changes do not break existing functionality. Whenever the model or related code is updated, the entire suite can be rerun to check for regressions.

- **Continuous integration / continuous deployment (CI/CD)**: Automated tests are a key part of CI/CD pipelines. When integrated into these pipelines, the tests can be triggered automatically whenever changes are pushed to the code base.

- **Speed of development**: By quickly identifying issues, automated test suites enable faster iteration and development of models, allowing teams to be more agile and responsive to changes.

- **Error reduction**: Manual testing is prone to human error, but automated tests perform the same steps precisely every time, reducing the chance of oversight or mistakes.

- **Documentation**: They serve as a form of documentation, showing new team members or stakeholders how the system is supposed to work.

- **Tooling**: There are various tools and frameworks available to help develop automated test suites. For example, in the Python ecosystem, `pytest` and `unittest` are popular for writing test cases, while Selenium can be used for browser-based tests if the model has a web interface.

Its implementation entails the following steps:

1. Define test cases that cover a full range of scenarios, including typical use cases, error handling, and performance benchmarks.

2. Write test scripts using a testing framework that's compatible with the technology stack of the LLM.

3. Set up a test environment that closely mirrors the production environment to ensure accurate results.

4. Integrate the test suite into the development workflow so that it runs automatically at key points, such as before merging code into the main branch.

5. Monitor the test results and maintain the test suite, updating it as the system evolves and new features are added.

Automated test suites are essential for maintaining the health and performance of LLMs throughout their development life cycle, from initial development through to maintenance and updates post-deployment.

Example of automated test suites in action

Consider a development team working on an LLM for customer support. To ensure the model functions correctly, they implement an automated test suite. Here are the attributes of an automated test suite:

- **Efficiency**: Thousands of test cases, including various customer queries, run automatically overnight, verifying performance across a range of scenarios

- **Consistency**: The suite reruns tests with every code update, ensuring that any changes do not introduce new issues

- **Comprehensiveness**: Edge cases, such as ambiguous language, are included, ensuring the LLM handles real-world situations effectively

- **Integration testing**: The suite tests how the LLM integrates with a backend database and frontend interface, ensuring seamless operation

- **Regression testing**: The suite ensures new features don't break any existing functionality, allowing safe updates

- **CI/CD integration**: The suite is part of the CI/CD pipeline, automatically testing every new code push to prevent issues from reaching production

- **Speed of development**: The suite enables rapid iteration by quickly identifying issues, allowing faster development and deployment

- **Error reduction**: Automated testing removes human error, ensuring accuracy every time tests are run

- **Documentation**: The test cases also act as documentation, helping new team members understand the LLM's expected behavior

- **Tooling**: The team uses `pytest`, `unittest`, and Selenium to write and execute tests, ensuring both backend and frontend functionality

By implementing this automated test suite, the team maintains the LLM's reliability and performance throughout development, enabling efficient and confident deployment.

Continuous integration

The practice known as **continuous integration (CI)** involves developers regularly incorporating their modifications into a unified code base. After this integration, the system automatically carries out testing and building processes. The primary intentions behind employing CI include enhancing the speed at which software defects are detected and corrected, boosting the overall caliber of the software, and minimizing the period necessary to approve and distribute updates to the software. Here's a detailed look at how CI is implemented and why it's beneficial, particularly for projects involving LLMs:

- **Automated builds**: Every time code is checked into the repository, the CI system automatically runs a build process to ensure that the code compiles and packages correctly. For LLMs, this might involve not just compiling code but also setting up the necessary data pipelines and environment for the model to run.

- **Automated testing**: After the build, the system executes a suite of automated tests against it. This could include unit tests, integration tests, and any other relevant automated tests that verify the functionality of the model and the integrity of the code.

- **Early bug detection**: By running tests automatically on every change, CI helps in identifying issues early in the development cycle. This is crucial for LLMs, where issues can be complex and hard to diagnose. Early detection leads to easier and less costly fixes.

- **Frequent code integration**: CI encourages developers to integrate their code into the main branch of the repository often (at least daily). This reduces integration problems and allows teams to develop cohesive software more rapidly.

- **Feedback loop**: Developers receive immediate feedback on their code changes. If a build or test fails, the CI system alerts the team, often through email notifications or messages in a team chat application.

- **Documentation**: The CI process often includes generating documentation or reports that detail the outcome of each build and test cycle, which can be vital for tracking down when and where issues were introduced.

- **Quality assurance**: Continuous testing assures the quality of the software. In the case of LLMs, it ensures that the model's performance is continuously monitored and that any degradation is flagged immediately.

- **Deployment readiness**: CI can help ensure that the code is always in a deployable state, which is particularly important for LLMs being used in production environments since stability is crucial.

- **Tools for CI**: There are many CI tools available, such as Jenkins, Travis CI, GitLab CI, and GitHub Actions, that can be configured to handle the build and test workflows for projects involving LLMs.

Implementing CI involves setting up a server where the CI process runs and configuring the project's repository to communicate with this server. The server monitors the repository and triggers the CI pipeline whenever it detects changes to the code base. For LLMs, CI servers might need to be equipped with the necessary hardware resources, such as GPUs for model training and testing, to handle the resource-intensive tasks involved in working with such models.

In summary, CI is an integral part of modern software development practices, including those involving LLMs. It helps maintain a high standard of code quality, encourages collaboration and communication among team members, and ensures that software products are always ready for deployment.

Example of a CI setup

Here's a very simple example of a CI setup using GitHub Actions for a Python project:

- **Python code** (`main.py`): This contains two basic functions – `add()` and `subtract()`:

```
def add(a, b):
    return a + b

def subtract(a, b):
    return a - b
```

- **Unit test** (`test_main.py`): This tests the `add()` and `subtract()` functions using Python's `unittest` framework:

```
import unittest
from main import add, subtract

class TestMain(unittest.TestCase):
    def test_add(self):
        self.assertEqual(add(1, 2), 3)

    def test_subtract(self):
        self.assertEqual(subtract(2, 1), 1)

if __name__ == '__main__':
    unittest.main()
```

- **CI configuration** (`ci.yml`): Please refer to the configuration example at `https://dev.to/rachit1313/streamlining-development-with-github-actions-a-ci-adventure-2116`. This simple CI pipeline ensures that your code is automatically tested every time changes are made, helping to catch errors early in the development process.

Stress testing

Stress testing, in the context of LLMs, is a critical evaluation method that's used to determine how a system operates under extreme conditions. The primary goal of stress testing is to push the system to its limits to assess its robustness and identify any potential points of failure. Let's take a closer look at the components and importance of stress testing for LLMs:

- **High-load simulation**: Stress testing involves creating scenarios where the LLM is expected to process a much higher volume of requests than usual. This can reveal how the model and its underlying infrastructure cope with sudden spikes in demand, which could occur during peak usage times or due to unexpected surges in popularity.

- **Large and complex inputs**: The model is fed with unusually large or complex data inputs to test the bounds of its processing capabilities. For an LLM, this might involve intricate, lengthy, or highly nuanced text sequences that are more challenging to analyze and generate responses for.

- **Performance metrics: Key performance indicators** (KPIs) such as response time, throughput, and error rates are monitored during stress tests. These metrics help to quantify the model's performance under pressure and can highlight performance degradation that may not be apparent under normal conditions.

- **Resource utilization**: Stress testing also provides data on how efficiently the model uses computational resources such as CPU, memory, and GPU under heavy loads. This can inform decisions about scaling and optimizing resource allocation.

- **Recovery assessment**: Another aspect of stress testing is to see how well the system recovers from failures. Do any components crash under heavy load, and if so, how does the system handle such crashes? Can the system gracefully degrade its service rather than fail outright?

- **Scalability**: The results of stress tests can indicate whether the current system setup can scale to meet future demands. They can help in planning for additional resources or in making architectural changes to support scalability.

- **Endurance**: Sometimes, stress testing is extended over a longer period to test the endurance of the system, ensuring it can handle sustained heavy use without performance decay or increased error rates.

- **Identifying bottlenecks**: Stress tests can reveal bottlenecks in data processing pipelines and other system components that may become critical under high load conditions.

Stress testing is an integral part of ensuring that an LLM is production-ready. It allows organizations to proactively address issues before they impact users and to ensure that the model can deliver consistent performance, even when pushed beyond typical operational expectations.

A/B testing

A/B testing, also known as split testing, is a method that's used for comparing two or more versions of a model or algorithm to determine which one performs better. It's a critical process in the development and refinement of LLMs and other AI systems. Here's an in-depth explanation of A/B testing and its relevance to LLMs:

- **Objective**: The primary goal of A/B testing is to make data-driven decisions based on the performance of different models. It involves exposing a similar audience to two variants (A and B) and using statistical analysis to determine which variant performs better based on specific metrics.

- **Randomization**: Requests are randomly assigned to either the control group (usually the current model) or the treatment group (the new or modified model) to eliminate any bias in the distribution of inputs that could affect the outcome of the test.

- **Metrics**: A/B testing for LLMs typically focuses on metrics that measure the quality and effectiveness of the model's outputs. This might include accuracy, response time, user engagement metrics, conversion rates, error rates, or any other relevant KPIs.

- **Segmentation**: Sometimes, A/B tests are conducted on specific segments of users to understand how different groups respond to the models. For instance, you could segment by demographic factors, user behavior, or even by the type of request being made.

- **Statistical significance**: It's essential to run the test until the results reach statistical significance, meaning that the outcomes that are observed are likely not due to chance. This typically requires a sufficient number of samples to ensure confidence in the results.

- **User experience**: In addition to objective performance metrics, A/B testing can also measure subjective aspects of user experience. Feedback can be collected directly from users or inferred from user behavior data.

- **Ethics and transparency**: When conducting A/B tests, it's important to maintain ethical standards and transparency, particularly if the test could impact the user experience. Users' privacy should be protected, and any changes to the user experience should be made with consideration of their potential impact.

- **Implementation**: To conduct an A/B test, you will typically need an A/B testing framework or platform that can route requests, collect data, and analyze results.

- **Iterative process**: A/B testing is often iterative. After analyzing the results of one test, the next iteration may involve refining the models based on insights gained and then testing again.

- **Decision-making**: The results of A/B tests are used to make decisions about whether to roll out a new model, continue developing and refining the model, or revert to the previous version.

A/B testing is a powerful technique for improving the performance of LLMs by allowing data-driven decisions to be made about which models best meet the needs of the users and the goals of the system. It's a user-centric approach that helps to ensure that models provide value and a positive experience.

Regression testing

Regression testing is a type of software testing that ensures that recent program or code changes have not adversely affected existing features. It's an essential component of quality assurance for software, including LLMs. Let's take a closer look at regression testing in the context of LLMs:

- **Purpose**: The main goal of regression testing is to confirm that the behavior and performance of an LLM remain consistent after modifications, such as updates to the code, model architecture, or training data.

- **Test cases**: A set of established test cases that the model has previously passed must be re-run. These test cases are typically automated and cover the full spectrum of the model's functionalities.

- **Scope**: The scope of regression testing can vary. In some cases, a small change may only require a subset of tests to be run (this is known as selective regression testing). In other cases, particularly for significant updates or over longer development cycles, the entire test suite may be executed.

- **Frequency**: Regression tests are run frequently throughout the development cycle, particularly after each significant code commit, before merging branches, or before a new version of the model is released.

- **Continuous integration**: In modern software development practices, regression tests are often integrated into a continuous integration pipeline, where they are automatically triggered by new code submissions.

- **Change impact analysis**: Part of regression testing is determining the impact of changes. If the changes are minor, the testing can be more targeted. For more significant changes, a comprehensive set of tests may be necessary.

- **Prioritization**: Sometimes, due to time constraints, it's necessary to prioritize which regression tests to run. Test cases that cover the most critical features of the LLM, or those that are most likely to be affected by recent changes, are run first.

- **Test maintenance**: As the LLM evolves, the regression test suite itself may need to be updated. New tests might be added, and obsolete tests might be removed to ensure the suite remains relevant and effective.

- **Results analysis**: The results of regression tests are analyzed to detect any failures. When a test case that previously passed now fails, it's an indication that a recent change may have introduced a bug.

- **Bug fixes**: If regression testing identifies a problem, the issue is fixed, and the test suite is run again to confirm that the fix is successful and hasn't caused any further issues.

- **Evaluation metrics**: Use appropriate evaluation metrics, both quantitative and qualitative, to measure the model's performance across the test cases. These metrics should align with the goals of the model and the needs of the end users.

Regression testing is crucial for maintaining the stability and reliability of an LLM over time. It helps developers and engineers ensure that improvements to the model do not come at the cost of previously established functionality and performance.

Version control

Version control acts as a tool that logs alterations to a file or group of files through time, allowing specific versions to be restored later. In the context of LLMs and their associated datasets, version control is essential for several reasons:

- **Reproducibility**: By maintaining version control over both the model's code base and the datasets used for training and testing, you can ensure that experiments are reproducible. This means that other researchers or developers can replicate your results, which is a cornerstone of scientific research and robust software engineering practices.

- **Traceability**: When issues arise, version control allows you to trace back and understand which changes might have introduced the problem. This is crucial for debugging and maintaining the integrity of the LLM.

- **Collaboration**: Version control systems such as Git facilitate collaboration among teams. Team members can work on different features or experiments in parallel, merge changes, and resolve conflicts in a controlled and transparent manner.

- **Documentation**: Version control also acts as a form of documentation. Commit messages and logs provide a history of the changes, why they were made, and by whom, which is invaluable for understanding the evolution of the model and its datasets.

- **Branching and merging**: Version control allows you to branch off from the main line of development to experiment with new ideas in a controlled environment. If these experiments are successful, they can be merged back into the main branch. If not, they can be discarded without affecting the main project.

- **Release management**: It helps in managing releases. You can tag specific commits that represent official releases or stable versions of the LLM, which is essential for deployment and distribution.

- **Model versioning**: Just like software, LLMs can be versioned. This is important because models may change over time due to them being retrained on new data or modifications being made to their architecture. Versioning ensures that the specific model used for any given task is identifiable.

- **Dataset versioning**: Datasets used in training and testing LLMs also change over time. Version control for datasets ensures that you know exactly which version of the data was used for each experiment, which is critical for replicating results and for the scientific integrity of the work.

Implementing version control effectively requires regular commits with clear, descriptive messages, tagging releases, branching for new features or experiments, and, perhaps most importantly, a culture of documentation and communication within the team. Tools such as Git, along with hosting services such as GitHub, GitLab, or Bitbucket, are commonly used to manage version control in software development and data science projects.

User testing

User testing is a crucial phase in the development cycle of any application, including those powered by LLMs. It involves real-world users interacting with the application to provide direct feedback on its performance and usability. Let's take an in-depth look at the role of user testing:

- **Real-world feedback**: Users often reveal practical issues and opportunities for improvement that developers may not have anticipated. User testing provides a reality check and ensures that the model meets the needs and expectations of its intended audience.

- **Usability and experience**: Through user testing, you can evaluate how intuitive and user-friendly the application is. This includes the ease with which users can complete tasks and how satisfying they find the interaction with the model.

- **Diversity of interaction**: Different users have unique ways of interacting with applications. User testing allows for a diverse range of interactions, which can uncover a wider array of issues or use cases that the LLM needs to handle.

- **Performance assessment**: While quantitative metrics can provide some insights into how well an LLM performs, user testing can evaluate subjective performance aspects, such as the relevance and helpfulness of the model's responses.

- **Contextual usage**: Users provide context for how the LLM will be used in day-to-day scenarios. They can offer valuable insights into how the model fits into real-life workflows and tasks.

- **Feedback loop**: User testing establishes a direct feedback loop for the development team. This information can be instrumental in prioritizing development tasks, fixing bugs, and iterating on the model's features.

- **Identification of edge cases**: Users may use the system in ways that developers didn't foresee, highlighting edge cases that need to be addressed to improve the robustness of the LLM.

- **Sentiment analysis**: Observing users' reactions can also provide qualitative data on the sentiment and emotional responses elicited by the LLM, which can be important for applications such as chatbots or virtual assistants.

- **Training data enrichment**: Interactions from user testing can sometimes be used to further train and refine the LLM, provided that privacy and data usage considerations are strictly adhered to.

- **Ethical and accessibility considerations**: User testing can also shed light on ethical considerations and accessibility issues, ensuring that the LLM is equitable and can be used by people with a wide range of abilities.

When conducting user testing, it's important to do the following:

- **Select a representative sample**: Users should represent the target demographic of the application

- **Ensure privacy**: Protect user data and ensure that the testing complies with all relevant privacy laws and regulations

- **Provide clear instructions**: Users should understand what is expected of them during the testing process

- **Collect structured feedback**: Use surveys, interviews, and analytics to collect and organize user feedback

- **Iterate**: Use the results from user testing to make iterative improvements to the model

User testing is an indispensable part of developing user-centric LLM applications, providing insights that cannot be captured through automated testing alone. It helps ensure that the final product is not only functional but also aligns well with user needs and preferences.

Ethical and bias testing

Ethical and bias testing are critical components in developing and deploying LLMs. This form of testing aims to identify and mitigate potential biases in the model's outputs and ensure that ethical standards are upheld. Let's take a detailed look at what this process entails:

- **Bias detection**:

  - Testing for bias involves evaluating the model's outputs for patterns that may indicate unfair or prejudiced treatment of certain groups. This can be based on race, gender, ethnicity, age, sexuality, or any other demographic factor.

  - Specialized test datasets that reflect a diverse range of identities and scenarios are used to probe the model's behavior and to uncover biases that might not be evident in more general datasets.

- **Ethical considerations**:

  - Ethical testing is conducted, which examines whether the model's outputs adhere to societal norms and values. It includes assessing the model for the potential to generate harmful, offensive, or inappropriate content.

  - This may also involve ensuring that the model respects user privacy and does not inadvertently reveal personal data.

- **Curated datasets for testing**: These are used for ethical and bias testing:

  - Datasets for ethical and bias testing are often carefully curated so that they include examples that challenge the model on ethical grounds or expose it to a wide variety of linguistic contexts related to sensitive issues

  - These datasets can be sourced from or inspired by real-world examples where bias has been an issue in the past, or they can be constructed by experts in ethics and social science to cover potential ethical dilemmas

- **Automated and manual evaluation**: Both are crucial for ethical and bias testing:

  - While some aspects of ethical and bias testing can be automated, manual review by human evaluators is essential. Diverse teams of reviewers can provide a range of perspectives that are invaluable for this type of testing.

  - Human evaluators can also assess the subtleties and nuances of language that automated systems may overlook.

- **Continuous monitoring**: This is very important:

  - Ethical and bias testing is not a one-time process. It requires ongoing monitoring and re-evaluation, especially as the model is exposed to new data and as societal norms evolve.

  - Models can be subject to "drift" over time, where their outputs change as they interact with users and additional data. Continuous monitoring helps ensure that these changes do not lead to the introduction of new biases or ethical issues.

- **Mitigation strategies**:

  - When biases or ethical issues are detected, mitigation strategies are employed. These can include retraining the model with more balanced data, implementing algorithmic fairness techniques, or adjusting the model's decision-making processes.

  - In some cases, constraints or filters may be implemented to prevent certain types of problematic outputs.

- **Transparency and accountability**:

  - Part of ethical testing involves creating transparency around how the model works and what types of data it has been exposed to. This can help stakeholders understand the model's decision-making process and the potential limitations of its outputs.

  - Accountability structures should be put in place to address any issues that may arise from the model's outputs and to provide recourse for those affected.

Ethical and bias testing is an essential practice to ensure that LLMs are fair, equitable, and aligned with societal values. It's an area that often involves interdisciplinary collaboration, bringing together expertise from data science, social science, ethics, and law.

Documentation

Documentation is an integral aspect of the testing process, serving as a record that outlines how testing is conducted, why certain decisions are made, and what the outcomes are. It's critical for ensuring transparency, facilitating future maintenance, aiding in knowledge transfer, and providing evidence for compliance with standards and regulations. Let's take an in-depth look at the various components and significance of documentation in the context of testing protocols for LLMs and other complex systems:

- **Test case documentation:**

  - Detailed descriptions of each test case, including the purpose, input conditions, execution steps, expected outcomes, and actual outcomes

  - Information on how test cases map to specific requirements or features of the LLM to ensure coverage of all functionalities

- **Testing process documentation:**

  - A comprehensive description of the testing methodology, including the types of testing performed (unit, integration, system, regression, and so on)

  - The rationale behind the chosen testing approach and methodologies, explaining why they are suitable for the LLM under test

- **Tools and environment:**

  - A list of the tools and technologies used in the testing process, such as testing frameworks, version control systems, continuous integration pipelines, and any specialized software for performance or security testing

  - Descriptions of the setup and configuration of the testing environment, including hardware specifications, operating systems, network configurations, and any other relevant infrastructure details

- **Results and reports:**

  - Test results, including pass/fail statuses for each test case, metrics collected (for example, response times, accuracy, and error rates), and any incidents or defects discovered

  - Summary reports and detailed logs that capture the outcomes of testing sessions, making it easier to track progress over time and to perform analyses if issues are detected

- **Version control**:

  - Documentation should be kept under version control, ensuring that changes to the testing documentation are tracked and that the history of updates is preserved

  - Links or references to the specific versions of the LLM and datasets used during testing, maintaining traceability between test results and the state of the system at the time of testing

- **Quality assurance and compliance**:

  - Evidence that testing protocols adhere to internal quality standards and any external regulations or industry standards that are applicable

  - Documentation of any quality assurance reviews, audits, or compliance checks that the testing protocols have undergone

- **Best practices and lessons learned**:

  - Insights gained from the testing process, including challenges encountered and how they were overcome, can inform future testing strategies

  - Best practices developed as part of the testing process can be standardized and applied to future projects

- **Maintenance and updates**:

  - Procedures can be implemented for updating and maintaining the testing documentation, ensuring it remains current as the LLM and its associated systems evolve

  - Plans for future testing cycles can be created, including any scheduled re-testing or plans for expanding the testing protocols as new features are added to the LLM

Proper documentation is not just a formality but a vital asset that supports the integrity and reliability of the LLM. It enables teams to work more effectively, provides a basis for decision-making, and ensures accountability throughout the model's life cycle.

Legal and compliance checks

Legal and compliance checks are vital processes within the testing protocols for LLMs to ensure that the model and its use comply with all applicable laws, regulations, and industry standards. Let's take a closer look at the aspects involved in legal and compliance checks:

- **Data privacy**: One of the most critical areas is data privacy. LLMs often require large datasets for training and testing, which may contain sensitive personal information. Legal and compliance checks ensure that any data that's handled adheres to privacy laws such as the **General Data Protection Regulation (GDPR)** in Europe, the **California Consumer Privacy Act (CCPA)**, or other relevant legislation.

- **User protection**: The model should be tested to ensure that it does not produce harmful outputs that could lead to user exploitation or harm. This includes checks against generating defamatory, libelous, or other types of illegal content.

- **Intellectual property**: Compliance checks involve verifying that the data used for training and testing the model does not infringe upon intellectual property rights. This means obtaining proper licenses for any copyrighted material included in the datasets.

- **Record-keeping**: Testing protocols must include rigorous record-keeping practices to document compliance with legal and ethical standards. This documentation can be crucial for demonstrating compliance in case of audits or legal inquiries.

- **Ethical standards**: Beyond legal requirements, LLMs should also adhere to ethical standards that may be set by industry bodies or the organization's ethical guidelines. This might involve issues such as fairness, transparency, and accountability.

- **Bias and fairness**: Legal and compliance checks should include assessments for bias and fairness, ensuring that the model does not exhibit unfair biases against certain groups, which could lead to discriminatory outcomes.

- **Accessibility**: Compliance with laws and regulations regarding accessibility, ensuring that the model is usable by people with disabilities, is also a crucial check. This could include compliance with the **Americans with Disabilities Act** (**ADA**) or the **Web Content Accessibility Guidelines** (**WCAG**).

- **Security**: The model and its data should be protected against unauthorized access and breaches. Compliance checks should verify that security measures are in place and align with industry standards such as ISO/IEC 27001.

- **International compliance**: For LLMs used across different regions, it's important to comply with international laws and regulations. This could involve additional complexity due to the variance in legal requirements from one country to another.

- **Continuous monitoring**: Legal and compliance requirements can change, so it's important to continuously monitor for any updates to the laws and regulations and adjust the testing protocols accordingly.

- **Consulting legal experts**: Involving legal counsel or compliance experts in the testing process can help you identify potential legal issues and develop strategies to address them. They can provide guidance on complex legal matters and help navigate the regulatory landscape.

Conducting thorough legal and compliance checks is essential not just for avoiding legal repercussions but also for building trust with users and stakeholders, as well as for ensuring the responsible development and deployment of LLMs.

Another aspect of testing is configuring and utilizing feedback loops, something we already covered in *Chapter 2, How LLMs Make Decisions*.

Human-in-the-loop – incorporating human judgment in evaluation

HITL is a concept where human judgment is used in conjunction with AI systems to improve the overall decision-making process. This integration of human oversight into the evaluation phase is particularly important for complex systems such as LLMs, where nuanced understanding and context may be required. Let's take a closer look at HITL in the context of LLM evaluation:

- **Enhanced decision-making**: Humans can provide nuanced assessments that go beyond what can be measured through automated metrics alone. This is especially critical for subjective areas such as language subtleties, cultural context, and emotional tone.

- **Quality control**: Involving humans in the evaluation process can help maintain high quality and accuracy in the model's outputs. Humans can catch errors or biases that automated tests might miss.

- **Training data refinement**: Human evaluators can help refine training data by providing feedback on the appropriateness and quality of the dataset, potentially identifying gaps or inconsistencies.

- **Model feedback**: By incorporating human feedback directly into the model's learning process, the LLM can be fine-tuned and improved. This feedback can come from evaluators, end users, or domain experts.

- **Interpretability and explainability**: Humans can help interpret the model's behavior and provide explanations for its outputs, which is essential for building trust and understanding among users.

- **Ethical oversight**: Human judgment is crucial when it comes to ethical considerations. Humans can ensure that the model aligns with ethical guidelines and social norms.

- **Continuous learning**: HITL systems can continuously learn from human inputs, leading to incremental improvements over time. This is a form of active learning where the model adapts based on human interactions.

- **Balancing automation and human insight**: Finding the right balance between automated evaluation and human judgment is crucial. While automation can handle a large volume of evaluation tasks, human insight is necessary for depth and context.

In practice, HITL can involve a range of activities, from annotating data, reviewing model outputs, providing qualitative feedback, and making judgments on the acceptability of the LLM's responses. The HITL approach ensures that LLMs are not only technically proficient but also practically useful and socially acceptable.

Ethical considerations and bias migration

The terms *ethical considerations* and *bias mitigation* are fundamental aspects of designing, developing, and deploying LLMs responsibly. Here's what each of these terms broadly encompasses within the context of AI and ML:

- **Ethical considerations**: This encompasses a wide array of principles and practices aimed at ensuring that LLMs behave in ways that are considered morally acceptable and beneficial to society. It involves the following aspects:

  - **Respect for privacy**: Ensuring that the LLM does not infringe on individuals' privacy rights and complies with data protection regulations

  - **Transparency**: Making the functioning of the LLM understandable to users, and clearly explaining the model's capabilities and limitations

  - **Accountability**: Establishing clear lines of responsibility for the outcomes produced by the LLM, including a framework for addressing any harm caused by the model's actions

  - **Fairness**: Ensuring that the LLM does not perpetuate or amplify biases and that it treats all users and groups equitably

  - **Non-maleficence**: Following the principle of "do no harm," ensuring that the LLM does not cause negative consequences for individuals or society

  - **Inclusivity**: Designing the LLM so that it's accessible to a diverse user base while considering factors such as language, abilities, and cultural backgrounds

- **Bias mitigation**: Bias in LLMs refers to systematic errors that unfairly discriminate against certain individuals or groups. Bias mitigation involves the following aspects:

  - **Identifying biases**: Using techniques to detect biases in data and model predictions, often requiring diverse and inclusive teams to recognize a broader range of potential biases

  - **Data correction**: Adjusting the training datasets to represent all pertinent demographics fairly, removing or reducing biased data points, and supplementing the data with more inclusive examples

  - **Algorithmic adjustments**: Tweaking the algorithms and model architectures to reduce the impact of biased data and prevent the model from learning these biases

  - **Continuous monitoring**: Regularly checking the model's outputs to ensure biases are not present or emerging as the model interacts with new data and users

  - **User feedback**: Incorporating feedback mechanisms for users to report biased or unfair outcomes, which can then be used to improve the model

  - **Impact assessment**: Evaluating the real-world impact of LLMs, especially on vulnerable or marginalized groups, to ensure the technology is being used ethically

Both ethical considerations and bias mitigation are ongoing processes. They require continuous attention and adaptation as societal norms evolve, new data is incorporated, and the LLM is put to use in varied contexts. Implementing robust ethical guidelines and bias mitigation strategies is essential for maintaining the trust of users and the public and for ensuring that the benefits of LLMs are realized without causing inadvertent harm or injustice.

Summary

Testing and evaluating LLMs is a multifaceted process that involves both quantitative and qualitative assessments to ensure their effectiveness and adherence to ethical standards. This critical phase goes beyond mere performance metrics; it includes human judgment through HITL evaluation methods to discern nuances that automated metrics may overlook. Additionally, it encompasses rigorous testing protocols to cover a wide spectrum of cases – from typical scenarios to edge cases and stress conditions – ensuring the LLM's robustness and readiness for real-world applications. Ethical considerations and bias mitigation are paramount, requiring continuous vigilance to ensure that the models act fairly and do not perpetuate existing prejudices. Through a combination of performance metrics, human evaluative input, and ethical oversight, this chapter aimed to help you establish LLMs that are not only high-performing but also equitable and responsible.

In the next chapter, we'll discuss the practice of deploying LLMs in production.

Part 3:
Deployment and Enhancing LLM Performance

This part addresses deployment strategies for LLMS, scalability and infrastructure considerations, security best practices for LLM integration, and continuous monitoring and maintenance. It also explains the alignment of LLMs with current systems, as well as seamless integration techniques, the customization of LLMs for specific system requirements, and security and privacy concerns in integration. Additionally, you will learn about quantization, pruning, and knowledge distillation, as well as advanced hardware acceleration techniques, efficient data representation and storage, how to speed up inference without compromising quality, and how to balance cost and performance in LLM deployment.

This part contains the following chapters:

- *Chapter 7, Deploying LLMs in Production*
- *Chapter 8, Strategies for Integrating LLMs*
- *Chapter 9, Optimization Techniques for Performance*
- *Chapter 10, Advanced Optimization and Efficiency*

7

Deploying LLMs in Production

Transitioning from theory to practice, in this chapter, we will address the real-world application of LLMs. You will learn about the strategic deployment of these models, including tackling scalability and infrastructure concerns, ensuring robust security practices, and the crucial role of ongoing monitoring and maintenance to ensure that deployed models remain reliable and efficient.

In this chapter, we're going to cover the following main topics:

- Deployment strategies for LLMs

- Scalability and infrastructure considerations

- Security best practices for LLM integration

- Continuous monitoring and maintenance

By the end of this chapter, you should be equipped with practical knowledge for transitioning from theory to the real-world application of LLMs.

Deployment strategies for LLMs

Choosing the right LLM for your specific application is a decision that can significantly affect the performance and outcomes of your system. Let's go through some detailed considerations to be taken into account.

Choosing the right model

When choosing the right model for your application, several key factors must be considered to ensure optimal performance and suitability for your specific needs. These factors include the following:

- **Model size**:

  - The size of an LLM, often denoted by the number of parameters it has, can range from millions to hundreds of billions. Larger models tend to have a better understanding of language nuances but are more computationally intensive and expensive to run.

  - Smaller models are more efficient and cost-effective but may not perform as well on complex language tasks. The choice of model size should balance the cost of operation against the required linguistic performance.

- **Language capabilities**:

  - LLMs vary in their ability to understand and generate text across different languages. Some models are trained primarily on English data, while others support multiple languages.

  - If your application targets a global audience or specific non-English speaking regions, it's important to choose a model with robust multilingual capabilities.

- **Learning approach**:

  - **Supervised learning**: These models are trained on labeled datasets and are excellent for tasks where the correct answers are known during training, such as classification problems.

  - **Unsupervised learning**: LLMs that use unsupervised learning can infer patterns from unlabeled data. They are useful for exploratory analysis, clustering, and generative tasks.

  - **Reinforcement learning**: LLMs trained with reinforcement learning improve their performance based on feedback from their environment. This approach is suitable for applications that involve a sequence of decisions, such as gaming or conversational agents that adapt to user preferences over time.

- **Domain-specific requirements**:

  - Certain applications may require a model that has been fine-tuned on domain-specific data. For instance, legal or medical applications would benefit from an LLM trained on relevant texts from those fields to understand the jargon and context better.

- **Ethical considerations**:

  - It's important to consider the ethical implications of deploying LLMs, especially regarding biases in the training data that could perpetuate stereotypes or discriminate against certain groups.

- **Vendor and community support**:

  - The choice of an LLM may also depend on the support offered by the vendor or the open source community. Having access to comprehensive documentation, active user communities, and reliable support can be critical for resolving issues during deployment.

- **Compliance and data governance**:

  - Depending on the region of deployment and the nature of the data being processed, different models may offer varying levels of compliance with data protection regulations such as GDPR or HIPAA.

- **Performance benchmarks**:

  - Before settling on a model, it's beneficial to evaluate it based on industry benchmarks or through proof-of-concept projects to assess its performance on tasks relevant to your application.

In summary, the decision to choose a particular LLM should be informed by a thorough understanding of your application's requirements and constraints. It's often recommended to perform pilot tests with different models to empirically determine which model performs best for your specific use case.

Integration approach

The integration of LLMs into existing systems is a critical step in leveraging their capabilities for real-world applications. The two primary methods for integrating LLMs are API integration and embedded integration, which we'll discuss next.

API integration

API integration, which involves connecting to an LLM through web-based service endpoints, offers numerous advantages, such as ease of use, simplified maintenance and upgrades, and cost-effectiveness. However, it also presents considerations and challenges. Let's review this further:

- **Definition and overview**:

  - **Application programming interface** (**API**) integration involves connecting to an LLM through web-based service endpoints. The LLM runs on external servers managed by the service provider, and the application interacts with it by sending HTTP requests and receiving responses.

- **Advantages**:

  - **Scalability**: API integration enables businesses to efficiently scale resources up or down based on demand, ensuring optimal resource utilization without over-provisioning.

- **Focus on core competencies (resource allocation)**: By utilizing API integration, companies can concentrate on their core strengths while outsourcing complex tasks such as machine learning model management.

- **Ease of use**: API integration is typically user-friendly, with well-documented endpoints that make it straightforward to send data and receive predictions.

- **Maintenance and upgrades**: The service provider is responsible for maintaining the model, ensuring it is up to date, and managing the underlying infrastructure.

- **Cost-effectiveness**: For applications with variable or low usage, this method can be cost-effective since you pay for what you use without investing in hardware.

- **Considerations and challenges**:

  - **Latency**: Each request to an API incurs network latency, which can be a bottleneck for applications requiring real-time processing.

  - **Dependence on internet connectivity**: API integration requires a reliable internet connection; any disruptions can lead to service unavailability.

  - **Data privacy**: Sending data to external servers may raise concerns about data security and privacy, particularly for sensitive information.

  - **Rate limiting**: APIs often have usage limits to prevent abuse, which could restrict the volume of requests your application can make.

  - **Limited customization of models**: The models provided by APIs are typically pre-trained and may offer limited customization options, potentially restricting their adaptability to specific business needs.

  - **No control on the quality**: Since the API provider controls the underlying models, businesses have no direct control over the quality or accuracy of the predictions, which can impact the overall reliability of the application.

  - **Vendor lock-in**: Relying heavily on a specific API provider can lead to vendor lock-in, making it challenging and costly to switch to a different service or provider in the future.

- **Use cases**:

  - API integration is ideal for applications that do not demand instantaneous responses and can tolerate some network latency, such as batch processing or asynchronous tasks.

Embedded integration

Embedded integration involves directly incorporating the LLM into the application's infrastructure, running it on the same servers or within the same environment. Let's explore it further:

- **Definition and overview**:

  - Embedded integration means incorporating the LLM directly into the application's infrastructure. The model runs on the same servers or within the same environment as the application.

- **Advantages**:

  - **Performance**: This approach minimizes latency since there are no external network calls, making it suitable for real-time applications.

  - **Data control**: Embedding the model locally allows for better control over data, which is critical for handling sensitive or proprietary information.

  - **Customization**: It offers the flexibility to customize the model and optimize it for specific tasks or performance requirements.

- **Considerations and challenges**:

  - **Resource intensive**: It requires significant computational resources, including powerful GPUs or TPUs, which can be expensive to acquire and maintain.

  - **Complex setup**: The setup is more complex and requires in-depth knowledge of **machine learning operations** (**MLOps**) to manage the model's life cycle effectively.

  - **Scalability**: Scaling an embedded model can be challenging and might require a sophisticated infrastructure setup with load balancing and auto-scaling capabilities.

- **Use cases**:

  - Embedded integration is well-suited for high-stakes or performance-critical applications, such as those in medical diagnostics, financial trading, or autonomous systems where low latency is paramount.

Choosing between API and embedded integration for deploying LLMs is a strategic decision that should align with the application's performance requirements, operational complexity, and resource allocation. Each approach has its own set of trade-offs and is best suited for different scenarios. Ultimately, the decision will depend on a thorough evaluation of the specific needs of your application, including technical requirements, data privacy concerns, and budgetary constraints.

Environment setup

Setting up the right environment to deploy LLMs is crucial for ensuring they operate efficiently and effectively. This setup involves a combination of hardware selection, software dependency management, and system compatibility checks. Here is a detailed breakdown of each component involved in the environment setup.

Hardware selection

When selecting hardware for LLMs, consider GPUs, which excel in parallel processing tasks and offer high computational speed, ample memory, and scalability for handling large models and datasets. Additionally, TPUs, optimized for ML workloads, are beneficial for training large models and offer cost-effectiveness in cloud environments.

As discussed previously, **GPUs** are specialized hardware designed to handle the parallel processing tasks that are common in ML and deep learning. They are highly efficient for both the training and inference phases of LLMs.

When selecting GPUs, consider the following:

- **Processing power**: Measured in **tera floating-point operations per second** (**TFLOPS**), which indicates the computational speed
- **Memory**: High **video random access memory** (**VRAM**), which is crucial for handling large models and datasets
- **Scalability**: The ability to scale horizontally by adding more GPUs if the workload increases
- **TPUs**: As custom chips developed specifically for ML workloads, they are optimized for the operations used in neural networks and can significantly accelerate the performance of LLMs

TPUs are particularly beneficial in the following cases:

- **Training large models**: They can speed up the training process by handling complex tensor operations efficiently
- **Improving cost-effectiveness**: In cloud environments, TPUs can offer a better price-to-performance ratio for certain workloads

While GPUs and TPUs handle the bulk of ML tasks, the CPU and system RAM are still important for the overall performance of the system

Ensure the CPU has enough cores and threads to efficiently handle the I/O operations, and there is sufficient RAM to support the overhead of the operating system and other applications

Software dependencies

When considering software dependencies for LLMs, ensure compatibility with the following:

- **Operating system**: Compatibility with the chosen operating system is essential. Most ML frameworks and tools are optimized for Unix-based systems, such as Linux distributions.

- **ML frameworks**: Frameworks such as TensorFlow, PyTorch, or JAX must be compatible with the hardware and have support for the specific model architectures you intend to use.

- **Libraries and drivers**: Install the necessary libraries and drivers that are compatible with your hardware. For GPUs, this includes **Compute Unified Device Architecture (CUDA)** for NVIDIA GPUs or ROCm for AMD GPUs.

- **Containerization**: Using containerization technologies such as Docker can help create consistent environments that are isolated from the rest of the system, simplifying dependency management and deployment.

System compatibility

When assessing system compatibility for LLM deployment, prioritize the following:

- **Integration with existing systems**: The environment should integrate seamlessly with your current infrastructure. This includes compatibility with data storage systems, networking configurations, and any other services your application relies on.

- **Version control**: Ensure that all software dependencies are version-controlled to avoid incompatibilities. Tools such as Git, along with package managers such as Conda or pip, can manage this.

- **Security protocols**: Implement security protocols that are compatible with your hardware and software stack to protect data and model integrity.

- **Monitoring and management tools**: Incorporate tools for monitoring the system's performance and managing resources. Examples include Prometheus for monitoring and Kubernetes for orchestrating containerized applications.

The environment setup for LLMs is a complex process that must be tailored to the specific needs of the application. It involves a careful balance of hardware capabilities, software dependencies, and system compatibility issues. By meticulously selecting the right components and ensuring they work harmoniously, organizations can create a robust and efficient environment that maximizes the performance of their LLMs.

Data pipeline integration

Before proceeding with data pipeline integration, it is essential for the user to thoroughly understand its objective and requirements. The objective typically involves ensuring that the data pipeline efficiently and accurately collects, processes, and delivers the necessary data to the LLM while meeting specific performance, scalability, and security standards. Key requirements may include data source identification, data quality benchmarks, processing speed, data privacy considerations, and the ability to scale with growing data volumes.

Integrating a robust data pipeline for LLMs is a multifaceted process, encompassing the collection, storage, preprocessing, and delivery of data to the model. An in-depth exploration of each stage in building a data pipeline for LLMs is as follows:

- **Data collection**:

  - **Data sources**: Identify diverse and reliable data sources that can provide the volume and variety of data required for LLMs. Sources can include websites, APIs, databases, and user-generated content.

  - **Data acquisition**: Establish mechanisms for acquiring data, such as web scraping, streaming data ingestion, or third-party data providers, while respecting data privacy and intellectual property laws.

  - **Data quality**: Implement quality checks to ensure the collected data is accurate, relevant, and unbiased. Poor data quality can lead to misleading model outcomes.

- **Data storage**:

  - **Choose between data lakes and warehouses**: This is dependent on the structure of your data and the need for scalability. Data lakes are suitable for storing raw, unstructured data, while warehouses are optimized for structured data.

  - **Scalability and accessibility**: The storage solution must be scalable to accommodate the ever-growing amount of data. It should also allow for easy retrieval and access when needed for training or inference.

  - **Data security**: Implement encryption, access controls, and other security measures to protect sensitive information and comply with regulations such as GDPR or HIPAA.

- **Data preprocessing**:

  - **Cleaning and normalization**: Raw data often contains noise and inconsistencies. Cleaning involves removing irrelevant or erroneous information, while normalization standardizes the data formats.

- **Tokenization and vectorization**: For language data, tokenization splits text into smaller units (tokens), and vectorization converts tokens into numerical representations that can be processed by LLMs.

  - **Feature engineering**: This involves creating data features that are particularly relevant to the task at hand, which can help improve model performance.

- **Data feeding**:

  - **Batching and buffering**: Organize data into batches for efficient processing and use buffering strategies to ensure a steady data flow to the model without overloading it.

  - **Data streaming**: For real-time applications, implement a data streaming mechanism that can continuously feed data into the LLM for instant inference.

  - **Data versioning**: Keep track of different versions of datasets to allow for reproducibility of results and to facilitate rollback in case of issues with new data.

Automation and orchestration

Automation and orchestration are an important part of data pipeline integration. The following techniques should be implemented:

- **Workflow management**: Use tools such as Apache Airflow or Luigi to automate and manage the data pipeline workflows, ensuring that the data processing steps are executed in the correct order

- **Continuous integration / continuous delivery (CI/CD)**: Implement CI/CD practices for the data pipeline to allow for continuous updates and deployment without disrupting the service

- **Monitoring and logging**: Establish comprehensive monitoring to track the health and performance of the data pipeline and set up logging to record events for debugging and audit purposes

A robust data pipeline is indispensable for the successful deployment of LLMs, as it ensures the consistent flow of high-quality data necessary for model training and inference. It requires careful planning, execution, and maintenance to address the challenges of big data management. By meticulously crafting each stage of the data pipeline, from collection to feeding, organizations can maximize the effectiveness of their LLMs, leading to improved outcomes, deeper insights, and more intelligent decision-making.

Scalability and deployment considerations

When deploying LLMs, considering scalability and infrastructure is crucial to ensure that the system can handle increased workloads without performance degradation. In this section, we will take a detailed look into the aspects of scalability and infrastructure considerations.

Hardware and computational resources

Setting up hardware and computational resources for LLM deployment is complex. Let's review them in detail in the following sections.

High-performance GPUs

GPUs, being the backbone of modern ML infrastructures due to their parallel processing capabilities, are ideal for the matrix and vector computations LLMs require.

When evaluating GPUs, consider the following:

- **Core count and speed**: A higher number of cores and faster clock speeds generally translate to better performance

- **Memory bandwidth and capacity**: Adequate memory is necessary to train large models, as it allows for larger batch sizes and faster data throughput

- **Scalability**: The ability to connect multiple GPUs can accelerate training and inference processes

Specialized AI processors (such as TPUs):

- TPUs, designed specifically for tensor computations, can provide faster and more energy-efficient processing for neural network tasks.

TPUs can be particularly useful for the following:

- **Distributed computing**: They are often optimized for parallel processing across multiple devices

- **Large-scale training**: TPUs can handle extensive computation loads, making them suitable for training very large models

High-performance CPUs

Although GPUs and TPUs handle the bulk of ML computations, CPUs are still important for general-purpose processing and orchestration tasks.

Look for CPUs with the following:

- **Multiple cores**: More cores mean better multitasking and parallel processing capabilities

- **High throughput**: Modern CPUs with high throughput can efficiently manage data pipelines and other I/O operations that are critical for LLMs

- **Networking**:

  - High-speed networking is essential for distributed training and data transfer between compute nodes

  - Implement low-latency networking hardware and software to ensure efficient communication, especially in clustered or cloud environments

- **Storage solutions**:

  - Fast and reliable storage solutions are necessary to store training data, model checkpoints, and logs

  - Consider SSDs for faster read/write speeds and high-capacity HDDs for long-term storage of large datasets

Infrastructure software

The following are important with regard to infrastructure:

- **ML frameworks**: Frameworks such as TensorFlow, PyTorch, and JAX should be optimized to leverage the underlying hardware, whether it's GPUs or TPUs

- **Distributed training libraries**: Libraries such as Horovod or TensorFlow's `tf.distribute` allow for scaling out the training process across multiple GPUs and machines

- **Orchestration and management tools**: Kubernetes for container orchestration and Terraform for infrastructure as code are vital for managing complex ML infrastructures

- **Monitoring and logging systems**: Implement systems such as Prometheus for monitoring and Grafana for visualization to keep track of the infrastructure's health and performance

Scalability strategies

When scaling LLM deployment, choose between the following:

- **Horizontal versus vertical scaling**:

  - Horizontal scaling involves adding more machines or nodes to the infrastructure, while vertical scaling means upgrading the existing machines with more power (for example, better CPUs or more memory)

  - Horizontal scaling is generally more flexible and robust for LLM workloads

- **Cloud-based versus on-premises solutions**:

  - Cloud services offer on-demand resource allocation and scalability without the need for significant upfront capital investment

  - On-premises solutions provide full control over the hardware and data, which might be required for compliance or security reasons

- **Elasticity and auto-scaling**: Implementing elastic resources that can be automatically scaled up or down based on the workload can optimize costs and performance

Infrastructure and scalability considerations form the foundation of a successful LLM deployment. It is not just about having the right hardware but also about how the infrastructure is designed to scale and adapt to changing demands. The goal is to balance performance with cost-effectiveness while ensuring that the system remains resilient and responsive as workloads grow. By planning for scalability from the outset, organizations can ensure their LLM deployments are future-proof and capable of supporting evolving ML tasks and applications.

Cloud versus on-premises solutions

The decision to utilize cloud-based services versus on-premises solutions for deploying LLMs is pivotal and depends on several factors including cost, control, compliance, and scalability. Both approaches have their own set of benefits and trade-offs that organizations must evaluate in the context of their specific needs.

Cloud-based services

The following items are relevant when it comes to cloud-based services:

- **Scalability**: Cloud services provide almost limitless scalability. Resources can be increased or decreased on demand, which is ideal for workloads that fluctuate over time.

- **Flexibility**: Users have the flexibility to choose from a variety of services and tools that cloud providers offer. This can include various types of storage, advanced analytics, and ML services.

- **Cost-effectiveness**: With a pay-as-you-go model, organizations only pay for the resources they use. This can be more cost-effective than investing in on-premises hardware that may not be used to its full potential.

- **Maintenance and upgrades**: The cloud provider is responsible for the maintenance and upgrade of the hardware and foundational software, which reduces the workload on internal IT teams.

- **Accessibility**: Cloud services can be accessed from anywhere, which is beneficial for remote teams or for businesses that operate in multiple locations.

- **Recovery and redundancy**: Cloud providers typically offer robust disaster recovery solutions and redundancy, which can be more sophisticated than what an organization might implement on-premises.

- **Disaster recovery**: Cloud services often include comprehensive disaster recovery options, ensuring that data can be quickly restored and operations can resume with minimal downtime in case of an unexpected event.

- **Access to advanced technologies**: Cloud providers regularly update their platforms with cutting-edge technologies, such as AI, big data analytics, and IoT services, allowing organizations to leverage the latest advancements without the need for significant internal investment.

On-premises solutions

Pay attention to the following regarding on-premises solutions:

- **Control**: On-premises infrastructure gives organizations full control over their hardware and software environment, which can be crucial for highly specialized or optimized LLM deployments.

- **Security**: Sensitive data remains on-site, which can be a significant advantage for organizations with strict data security requirements. There's a reduced risk of data breaches associated with external networks.

- **Compliance**: Certain industries have regulatory requirements that dictate how and where data is stored and processed. On-premises solutions can make it easier to comply with these regulations.

- **Performance**: On-premises solutions can offer better performance, especially if the organization has the resources to invest in high-end hardware and optimized networking solutions.

- **Cost predictability**: Although the initial investment is higher, on-premises solutions offer predictable costs over time, without the variability associated with cloud services.

- **Customization**: On-premises infrastructure can be highly customized to meet the specific needs of the organization, which can be important for specialized computing tasks.

Hybrid solutions

Many organizations opt for a hybrid approach, where some components are hosted on the cloud while others remain on-premises. This can offer a balance between the flexibility and scalability of cloud services and the control and security of on-premises solutions:

- **Data sovereignty**: A hybrid model can help navigate data sovereignty issues by keeping sensitive data on-premises while leveraging the cloud for computational tasks

- **Cost and performance optimization**: Organizations can optimize costs and performance by using the cloud for high-demand periods or specific tasks while maintaining an on-premises infrastructure for baseline workloads

- **Transition and scalability**: A hybrid approach allows for a gradual transition to the cloud, providing scalability as the organization's needs grow

Deciding between cloud-based services and on-premises solutions is a strategic decision that should consider the organization's specific needs, regulatory environment, and operational flexibility. The cloud offers scalability and cost-effective resource management, while on-premises solutions provide greater control and security. A thorough evaluation of both the long-term strategic goals and the operational capabilities of the organization will guide this decision, potentially leading to a combination of both in a hybrid model.

Load balancing and resource allocation

Load balancing and resource allocation are crucial components of managing a computational infrastructure, especially when it comes to deploying and operating LLMs. Here is a detailed look at both concepts.

Load balancing

Let's have an overview of load balancing:

- **Definition**: Load balancing distributes network or application traffic evenly across several servers or nodes to prevent any single one from becoming a bottleneck, ensuring system performance is maintained and outages are avoided

- **Methods**:

  - **Round robin**: Distributes requests sequentially across the servers in the pool

  - **Least connections**: Directs traffic to the server with the fewest active connections

  - **Resource-based**: Considers the current load and the capacity of each server to handle additional work

  - **Hybrid approaches or custom approaches**: Combines multiple load balancing strategies or tailors specific approaches to suit unique application requirements, providing more flexibility and optimization

  - **Dynamic load balancing**: Continuously monitors server performance and dynamically adjusts the distribution of traffic based on real-time data, ensuring optimal resource utilization

  - **Geographic load balancing**: Distributes traffic based on the geographic location of the user, routing them to the nearest or most efficient server to reduce latency and improve user experience

- **Considerations**:

  - **Session persistence**: Some applications may require session persistence, where consecutive requests from a single client are sent to the same server

  - **Health checks**: Regularly checking the health of servers to ensure traffic is not directed to failed nodes

  - **Scalability**: The load balancing solution itself must be scalable to adapt to the changing number of requests

- **Technologies**: Hardware load balancers, software-based solutions such as HAProxy, or cloud-based load balancers provided by services such as AWS Elastic Load Balancing

Resource allocation

Resource allocation involves assigning the available computational resources, such as CPU time, memory, and storage, to various tasks in a way that maximizes efficiency and prevents resource contention.

- **Strategies**:

  - **Static allocation**: Assigning fixed resources to specific tasks or services, which can be simple but may not be efficient

  - **Dynamic allocation**: Resources are allocated on the fly based on current demand and workload characteristics

  - **Resource pooling**: Consolidating resources into a shared pool that can be dynamically distributed across tasks or services as needed, improving resource utilization and flexibility

  - **Prioritization and queuing**: Implementing systems that prioritize tasks based on importance or urgency, with lower-priority tasks being queued for later processing, ensuring that critical operations receive the necessary resources first

- **Considerations**:

  - **Priority**: Some tasks may be more critical and require prioritized resource allocation

  - **Resource limits**: Preventing any single task from using an excessive amount of resources, which could starve other processes

  - **Resource reservation**: Reserving resources for high-priority tasks to ensure they can be handled immediately when they arise

- **Tools and technologies**: Container orchestration systems such as Kubernetes, which can automate resource allocation and provide fine-grained control over how resources are used by different containers

Combining load balancing with resource allocation

In the context of LLMs, combining load balancing with resource allocation can be particularly effective in the following:

- **Handling variable workloads**: LLMs may experience highly variable workloads, with periods of high demand followed by quieter times. Efficient load balancing and resource allocation can handle these fluctuations without over-provisioning.

- **Optimizing costs**: By balancing the load and allocating resources dynamically, organizations can optimize their infrastructure costs, paying more only when demand is high.

- **Ensuring high availability**: Distributing the load and managing resources effectively ensures that the LLMs are always available to handle requests, which is essential for services that require high uptime.

Load balancing and resource allocation are key to maintaining the responsiveness and reliability of systems that deploy LLMs. Effective strategies in these areas lead to improved performance, better resource utilization, and cost savings. They are particularly important as the complexity and scale of tasks for LLMs grow, requiring more sophisticated infrastructure management techniques to keep systems running smoothly.

Security best practices for LLM integration

To secure data privacy in LLM integrations, we can use encryption for data at rest and in transit, anonymize sensitive information, and enforce robust access controls. In this section, we will learn how to implement data minimization, secure sharing practices, and implement differential privacy. We will also go through the importance of regularly auditing for compliance, integrating security across the development life cycle, establishing firm data retention rules, and providing continual security training for staff.

Data privacy and protection

Ensuring the security of LLMs during integration into systems involves a comprehensive approach to data privacy and protection. Here are detailed best practices for securing LLM integrations:

- **Encryption**:

  - **At-rest encryption**: All sensitive data stored for LLM use should be encrypted. This includes training data, model parameters, and user data. Techniques such as **Advanced Encryption Standard** (**AES**) are commonly used for this purpose.

  - **In-transit encryption**: Data transmitted to or from LLMs should be protected using protocols such as **Transport Layer Security** (**TLS**) to prevent interception and unauthorized access.

- **Anonymization and pseudonymization**:

  - **Data anonymization**: Before feeding data into an LLM, remove all PII. Techniques such as data masking or tokenization can replace sensitive elements with non-sensitive equivalents.

  - **Pseudonymization**: This is a method where you replace private identifiers with fake identifiers or pseudonyms. This allows data to be matched with its source without revealing the actual source.

- **Access control**:

  - **Authentication**: Ensure that only authenticated users can access the LLM or the data it processes. This might include **multi-factor authentication** (**MFA**) mechanisms.

  - **Authorization**: Implement role-based access control to ensure that users have the minimum necessary permissions to perform their jobs.

- **Data minimization**: Collect and process only the data that is absolutely necessary for the LLM to perform its function. This not only reduces the risk of data breaches but also complies with data protection regulations such as GDPR.

- **Secure data sharing**: When sharing data between systems or with third parties, ensure that it is done securely and with the necessary legal agreements (such as NDAs) in place.

- **Differential privacy**: If the LLM's outputs are shared publicly, use differential privacy techniques to add noise to the data or the model's outputs, making it difficult to trace data back to any individual.

- **Regular audits and compliance checks**: Conduct regular security audits to ensure data privacy practices are up to date and effective. This includes compliance with legal standards and regulations.

- **Secure development life cycle**: Integrate security into the development life cycle of the LLM application. This involves security reviews at each stage of development, from design to deployment.

- **Data retention policies**: Establish and enforce data retention policies that dictate how long data is kept and when it should be securely deleted.

- **Training and awareness**: Regularly train staff on the importance of data privacy and the specific measures they must take to protect it. This includes training on recognizing phishing attempts and other security threats.

The integration of LLMs into any system requires a strong emphasis on data privacy and protection. By employing a combination of encryption, anonymization, access control, and adherence to privacy principles, organizations can significantly mitigate the risk of data breaches and unauthorized access. Continuous monitoring, regular audits, and a culture of security awareness are equally important to maintain a robust security posture.

Access control and authentication

Once authorizations are in place, access control and authentication can be determined. Access control and authentication are fundamental components of security frameworks, especially when it comes to protecting sensitive systems and data associated with LLMs. Let's go through an in-depth discussion of access control and authentication in the context of LLM integration.

Access control

The following are relevant regarding access control:

- **Role-based access control (RBAC):**

  - RBAC is a widely used approach where access rights are granted according to the roles of individual users within an organization. It ensures that users can only access the information that is necessary for their roles.

  - This approach simplifies the management of user permissions and can be easily updated as roles change or evolve within an organization.

- **Attribute-based access control (ABAC):**

  - ABAC uses policies that combine multiple attributes, which can include user attributes (role, department, and so on), resource attributes (owner, classification, and so on), and environmental attributes (time of day, location, and so on).

  - ABAC provides finer-grained control compared to RBAC and can dynamically adjust permissions based on a wide range of variables.

- **Access control lists (ACLs):**

  - ACLs are used to define which users or system processes are granted access to objects, as well as what operations are allowed on given objects.

  - In an ACL, each item outlines who can perform what action on a resource; for instance, it might allow John to have read access to `Report.txt`.

- **Mandatory access control (MAC):** In MAC, access rights are regulated based on fixed security attributes or labels. This model is often used in environments that require a high level of confidentiality and classification of data.

Authentication

Authentication encompasses the following:

- **Password-based authentication**:

    - The most common form of authentication involves verifying the identity of a user by validating their secret password

    - Password policies should enforce complexity requirements and expiration times, and prevent the reuse of passwords

- **MFA**:

    - MFA requires users to provide two or more verification factors to gain access to a resource, significantly increasing security

    - Factors can include something you know (password), something you have (a smartphone or hardware token), and something you are (biometrics)

- **Biometric authentication**:

    - Systems may use biometric methods such as fingerprint scans, facial recognition, or iris scans to authenticate users

    - While biometric authentication can be very secure, it also raises privacy concerns and requires careful handling of biometric data

- **Single sign-on (SSO)**: SSO allows users to authenticate once and gain access to multiple systems without re-authenticating. This is convenient for users and reduces the number of credentials that need to be managed.

- **Certificate-based authentication**: This method uses digital certificates to authenticate a user, machine, or device. The certificate is typically issued by a trusted **certificate authority (CA)** and is a form of **public key infrastructure (PKI)**.

Implementation considerations

To enhance security in LLM integration, we need to enforce strict access controls, use the principle of least privilege, regularly audit system access, segregate duties, and manage user accounts diligently. These measures prevent unauthorized access and maintain data integrity. Adopting the following comprehensive security measures is vital for the secure integration of LLMs into systems:

- **Least privilege principle**:

    - Users should be given the minimum levels of access—or permissions—needed to perform their job functions

  - This principle reduces the risk of an insider accidentally or maliciously accessing sensitive data or systems

- **Regular audits and reviews**: Regularly review access controls and authentication logs to ensure compliance with policies and to detect any irregularities or unauthorized access attempts.

- **Segregation of duties**: Critical functions should be divided among different individuals to prevent fraud or error. This is particularly important in financial or sensitive operations.

- **User account management**: Processes should be in place for creating, modifying, disabling, and deleting user accounts as part of the employee life cycle.

A robust security posture incorporating strict access control policies and strong authentication mechanisms is essential when integrating LLMs into any system. This ensures that only authorized personnel can access the LLM and its data, thereby maintaining the integrity and confidentiality of the system. By employing a combination of these strategies, organizations can protect themselves against a wide array of security risks, ensuring that their deployment of LLMs is as secure as possible.

Regular security audits

Regular security audits are a critical component of maintaining the integrity and trustworthiness of systems, especially those involving LLMs. Here's a detailed look into how regular security audits are conducted and why they are important.

Purpose of security audits

Security audits serve the following utilities:

- **Identification of vulnerabilities**: Audits systematically evaluate the security of a system's information by assessing how it conforms to a set of established criteria. They reveal weaknesses that could be exploited by threats.

- **Verification of compliance**: Regular audits check adherence to laws, regulations, and policies that govern data security and privacy, ensuring legal and regulatory compliance.

- **Risk assessment**: Audits help in identifying and prioritizing risks, allowing organizations to allocate resources effectively to mitigate these risks.

Conducting security audits

- **Planning**: Define the scope of the audit, objectives, and timelines. Decide whether the audit will be conducted internally, externally, or a combination of both.

- **Reviewing documentation**: Examine policies, procedures, and records. This includes access control policies, user account management protocols, and previous audit reports.

- **System and network scanning**: Use tools to scan for vulnerabilities. This may involve penetration testing, where the auditors simulate attacks to test the system's defenses.

- **Physical security checks**: Evaluate the physical access controls to the hardware and network components to ensure there are no physical vulnerabilities.

- **User access and privileges review**: Assess user permissions to ensure the principle of least privilege is being followed.

- **Data protection measures**: Verify that data encryption, anonymization, and backup strategies are properly implemented and effective.

Post-audit activities

- **Reporting**: Prepare a detailed audit report that outlines what was examined, what vulnerabilities were found, and recommendations for remediation.

- **Remediation**: Address the vulnerabilities identified in the audit report. This may involve patching software, updating policies, or enhancing security protocols.

- **Follow-up audits**: Conduct follow-up audits to ensure that the corrective actions have been implemented and are effective.

Types of security audits

- **Internal audits**: Conducted by the organization's own audit staff. They are beneficial for ongoing assurance and can be more cost-effective.

- **External audits**: Performed by independent organizations. They can provide an objective assessment and may be required for regulatory compliance.

- **Automated audits**: Utilizing software tools to regularly scan for vulnerabilities. While they can't replace comprehensive audits, they are useful for ongoing monitoring.

Best practices

- **Regular schedule**: Conduct audits at regular intervals, such as annually, or after any significant changes to the system or policies

- **Comprehensive coverage**: Ensure the audit covers all aspects of the system, including hardware, software, networks, and policies

- **Qualified auditors**: Use qualified personnel who have the necessary skills and knowledge to conduct thorough audits

- **Continuous improvement**: Use the findings from audits to continuously improve security practices

Regular security audits are essential for identifying vulnerabilities and ensuring compliance with security policies and regulations. They are a proactive measure that can prevent security breaches and instill confidence in the organization's commitment to protecting its assets and data. By incorporating regular security audits into their security strategy, organizations can significantly reduce their risk profile and respond more effectively to the evolving threat landscape.

Continuous monitoring and maintenance

Continuous monitoring and maintenance are pivotal practices in the life cycle of deploying LLMs. We will cover the specifics of these practices next.

Continuous monitoring

To ensure the effective operation of LLMs, monitor critical performance metrics such as model accuracy, response time, and error rates. System health should also be tracked, focusing on resource utilization, network performance, and service availability. Let's review them further:

- **Performance metrics**:

    - **Accuracy**: Regularly measure the model's prediction accuracy to ensure it is within acceptable thresholds for its intended application

    - **Response time**: Monitor the latency from when a request is made to the model to when a response is received, as excessive delays can impact user experience

    - **Error rates**: Track the rate of errors or unexpected outputs, which can signal issues with the model itself or the data it is processing

- **System health monitoring**:

    - **Resource utilization**: Keep an eye on CPU, GPU, memory, and disk usage to ensure the infrastructure is not overburdened

    - **Network performance**: Monitor network throughput and error rates to detect connectivity issues that could affect model performance

    - **Service availability**: Use uptime monitoring tools to ensure the LLM services are consistently available

    - **Task-specific parameter monitoring using dashboards**: Leverage dashboards to monitor specific parameters related to different tasks, providing a visual representation that allows for quick assessment and identification of any anomalies or performance issues

- **Automated alerts**: Implement an alerting system to notify relevant personnel when performance metrics fall outside of predefined thresholds

- **Monitoring tools**: Utilize comprehensive monitoring solutions such as Prometheus, Grafana, or the **Elasticsearch, Logstash, and Kibana** (**ELK**) stack for real-time data visualization and analysis

Maintenance practices

To ensure the ongoing efficacy and security of LLMs, it's vital to regularly retrain models with updated data, refine algorithms, and implement infrastructure and software enhancements. This maintenance strategy should also include rigorous compliance reviews, security updates, and effective backup and recovery systems. Here's an in-depth review:

- **Model retraining and updates**:

  - Periodically retrain the model with new data to maintain or improve its accuracy, especially as the nature of the input data evolves over time

  - Update the model to incorporate improvements in algorithms or to address discovered biases

- **Software updates**: Regularly update the software stack, including the operating system, ML frameworks, libraries, and dependencies, to patch security vulnerabilities and improve performance

- **Infrastructure upgrades**: Upgrade the underlying hardware and infrastructure as needed to handle increased loads or to improve computation speed

- **Data pipeline refinement**: Continuously improve the data pipeline to enhance data quality, address data drift, and ensure the pipeline's efficiency and reliability

- **Security patching**: Apply security patches promptly to protect against new vulnerabilities

- **Compliance checks**: Regularly review the system against compliance standards to ensure it meets all legal and regulatory requirements

- **Backup and recovery**: Maintain up-to-date backups of the LLM and its associated data and ensure that disaster recovery plans are in place and tested

- **Documentation and change management**: Keep detailed records of the system's configuration and changes over time to support maintenance activities and audits

Continuous monitoring and maintenance are essential for the long-term success and reliability of LLM deployments. They involve the ongoing assessment of performance metrics, system health, and user feedback, coupled with regular updates and improvements. By institutionalizing these practices, organizations can ensure that their LLMs continue to perform effectively, securely, and in compliance with relevant standards and regulations.

Summary

Deploying LLMs in production transitions from theoretical understanding to practical application, necessitating strategic planning to ensure the models' reliability and efficiency. The process involves careful consideration of deployment strategies that suit the application's needs, managing scalability and infrastructure to handle computational demands, and implementing robust security practices to safeguard sensitive information. Integral to the deployment is a regime of continuous monitoring and maintenance, which includes performance tracking and periodic updates or retraining of models to adapt to new data patterns and evolving user requirements. This chapter systematically covered these core aspects to equip you with the necessary insights for successful LLM integration and long-term operation.

In the next chapter, we will lay out the strategies for integrating LLMs.

8
Strategies for Integrating LLMs

Here, we will offer an insightful overview of integrating LLMs into existing systems. We will cover the evaluation of LLM compatibility with current technologies, followed by strategies for their seamless integration. We will also delve into the customization of LLMs to meet specific system needs and conclude with a critical discussion on ensuring security and privacy during the integration process. This concise guide will provide you with the essential knowledge to effectively incorporate LLM technology into established systems, while maintaining data integrity and system security.

In this chapter, we're going to cover the following main topics:

- Evaluating compatibility – aligning LLMs with current systems
- Seamless integration techniques
- Customizing LLMs for system-specific requirements
- Addressing security and privacy concerns in integration

By the end of this chapter, you will have gained a comprehensive understanding of integrating LLMs into existing systems.

Evaluating compatibility – aligning LLMs with current systems

Evaluating compatibility, which is ensuring that LLMs align with current systems, is a multifaceted task that requires a nuanced approach to technology integration. This process seeks to harmonize the capabilities of LLMs with the technological and operational fabric of an existing system, enhancing its functionalities without disrupting established workflows. Let's explore this intricate process in detail.

Technical specifications assessment

Navigating the complex landscape of technical specifications is paramount for effectively integrating LLMs into existing systems. Let's explore this further.

Computing power and storage

In the realm of optimizing computing power and storage infrastructure for LLMs, several key considerations come into play:

- **Computing power**:

  - **Processor requirements**: LLMs typically require high-performance processors to handle the complex computations involved in natural language processing. This often means using servers equipped with multi-core CPUs for parallel processing capabilities.

  - **GPU acceleration**: Many LLM operations, especially those involving neural networks, are significantly faster on GPUs than on traditional CPUs, due to the GPU's ability to handle thousands of threads simultaneously. The parallel processing capabilities of GPUs make them particularly well-suited for the matrix and vector operations that are common in machine learning.

  - **TPUs and other accelerators**: Some organizations may look beyond GPUs to TPUs and other specialized hardware accelerators designed explicitly for machine learning tasks, which can offer even higher efficiency for certain types of computations.

- **Storage**:

  - **Capacity needs**: LLMs not only require space to store a model but also need to accommodate the potentially vast amounts of data used for training and inference. This can include the datasets the model is trained on, any intermediate data created during processing, and the storage of multiple model versions.

  - **Data access speed**: The speed at which data can be read and written is also a consideration. **Solid state drives** (**SSDs**) or even faster storage solutions such as **Non-Volatile Memory Express** (**NVMe**) may be required to reduce data access latency, which is crucial for maintaining efficient processing times.

  - **Distributed storage systems**: For very large datasets or models, distributed storage systems that can scale horizontally, such as object storage solutions such as Amazon S3 or distributed file systems such as HDFS, might be necessary.

  - **Logs**: Adequate storage must be allocated for logs generated during model training, inference, and system operations, as they are essential for debugging, performance analysis, and compliance purposes.

  - **Configuration files**: Storing configuration files that define model parameters, environment settings, and operational controls is crucial for replicability, consistency across deployments, and efficient management of model versions.

  - **Output storage**: Ensure that there is sufficient space to store the outputs generated by a model, such as predictions, transformed data, or reports, particularly when dealing with large-scale batch processing or continuous data streams.

Assessing the current system

To ensure that the current system is capable of meeting the demanding requirements of LLMs, a comprehensive assessment of technical specifications is paramount. This involves evaluating infrastructure performance, considering upgrade paths, analyzing cost implications, ensuring compatibility and future-proofing, and prioritizing scalability for future growth. By meticulously examining these factors, organizations can equip themselves with the necessary resources and capabilities to effectively leverage the power of LLMs and unlock their full potential.

- **Infrastructure evaluation**:

  - **Benchmarking**: Current systems should be benchmarked to evaluate their performance capabilities. This includes running tests that simulate the workload of an LLM to see whether the processing and storage infrastructure can handle the load.

  - **Upgrade paths**: If the current infrastructure is insufficient, organizations need to consider their upgrade paths. This may involve investing in new hardware, migrating to a cloud-based solution that can offer scalable compute and storage resources, or adopting a hybrid approach.

- **Cost considerations**:

  - **Capital expenditure versus operational expenditure**: The decision to upgrade hardware (a capital expenditure) versus utilizing cloud services (an operational expenditure) involves not only a technical assessment but also a financial one. Cloud services can offer scalability and reduce the need for upfront investment, but over time, they may become more expensive than owning and operating your own hardware.

  - **Energy efficiency**: The energy consumption of LLMs, especially when using GPU or TPU accelerators, can be significant. It's essential to factor in not just the cost of the hardware but also the ongoing energy costs.

- **Compatibility and future-proofing**:

  - **System integration**: The new or upgraded hardware must be compatible with the existing infrastructure. This means considering the physical requirements (such as rack space in data centers), the compatibility with existing software and operating systems, and network requirements.

  - **Scalability for future needs**: When upgrading or choosing new infrastructure, it's important to consider not just the current needs but also future growth. Scalability ensures that an infrastructure can handle increasing loads without requiring a complete overhaul shortly.

In summary, evaluating the technical specifications thoroughly ensures that an existing system can handle the demands of LLMs. This process includes assessing current infrastructure, planning for future needs, balancing costs, and ensuring compatibility with existing technology.

Understanding data formats

Understanding data formats for LLMs is essential for seamless integration. Let's explore this in detail.

Understanding data formats for LLMs

LLMs typically require data in a structured format that can be systematically parsed and understood by a model. Common data formats for LLMs include the following:

- **JavaScript Object Notation (JSON)**: A data-interchange format that is lightweight and easy for machines to generate and parse, as well as for humans to read and write

- **Comma-separated Values (CSV)**: A simple format that stores tabular data, such as in databases or spreadsheets, where each record is on a line and fields within a record are separated by commas

- **TXT**: A plain text file that contains unformatted text, often used for models that require only textual data without additional metadata

- **eXtensible Markup Language (XML)**: A markup language that establishes rules for document encoding in a format that is readable by both machines and humans

Transforming data for compatibility

If an existing system does not natively output data in an LLM-compatible format, a transformation layer is required to convert the data into a suitable format. This involves several steps:

1. **Data extraction**: Extracting the necessary information from the source format. For instance, if the source data is in XML, this would involve parsing the XML to extract the relevant fields.

2. **Data transformation**: Converting the extracted data into the format required by the LLM. This could involve restructuring the data to fit a JSON schema, ensuring that all necessary attributes are present and that the format aligns with what the LLM expects.

3. **Data loading**: Loading the transformed data into the LLM's environment, which could be a database, or directly into the model for processing.

Tools and technologies for data transformation

The process of converting data formats can be facilitated by various tools and technologies, such as the following:

- **ETL (Extract, Transform, Load) tools**: Software such as Informatica, Talend, and Apache NiFi are specifically designed to handle the flow of data between systems and transform it as necessary.

- **Scripting languages**: Python or Perl scripts, often used due to their powerful text processing capabilities and libraries to handle different data formats.

- **Middleware**: Software that mediates between different systems or applications, providing services for data conversion and communication.

- **Application programming interfaces (APIs)**: APIs that can provide on-the-fly conversion services. For example, a REST API might accept data in the XML format and return it in the JSON format.

Best practices for data format conversion

- **Validation**: After conversion, validate the data to ensure that the transformation process has not introduced errors or corrupted it.

- **Logging**: Maintain logs of the transformation processes for debugging purposes and to ensure the traceability of data.

- **Performance optimization**: Since data transformation can be resource-intensive, it's important to optimize computing for performance, especially if dealing with large volumes of data. This might involve parallel processing or in-memory computations.

- **Error handling**: Implement robust error handling to manage issues that may arise during the transformation process, such as missing fields or incompatible data types.

- **Security**: Ensure that the data transformation process adheres to security best practices, especially when handling sensitive data.

- **Scalability**: The data transformation solution should be scalable and able to handle growing amounts of data without significant reengineering.

In summary, data format compatibility is a critical component in the integration of LLMs into existing systems. By establishing a reliable and efficient transformation layer, organizations can ensure that their data flows seamlessly from their native systems into the LLM, thereby leveraging the full capabilities of AI for their applications. This not only enhances the performance of the LLM but also ensures that the integration process is smooth and efficient, minimizing disruption to existing workflows.

Compatibility with programming languages, APIs, and frameworks

Integrating LLMs into existing systems requires careful consideration of compatibility with programming languages, APIs, and frameworks. Let's review this further.

Programming languages

The choice of programming languages and development environments plays a significant role in the integration of LLMs into existing systems. When the development environment is aligned with the requirements of the LLMs, integration becomes a more streamlined process. The following is a detailed exploration of how programming languages affect the integration of LLMs and considerations to keep in mind:

- **Programming languages and LLM integration**:

  - **Language compatibility**:

    - **Native SDK support**: LLM providers often supply **Software Development Kits (SDKs)** in popular programming languages such as Python, which simplify the process of integrating a model into existing systems. An SDK includes libraries, tools, and documentation that allow developers to work with an LLM more easily.

    - **Community and library support**: Languages with large developer communities and extensive libraries, such as Python, Java, and JavaScript, are often preferred because they offer pre-built modules and community support that can accelerate development and problem-solving.

    - **Performance considerations**: The chosen programming language should be capable of handling the performance requirements of an LLM. For example, Python is widely used for machine learning because of its simplicity and the rich ecosystem of data science libraries. However, for performance-critical sections of the code, languages like C++ can be used in conjunction with Python.

  - **Development environment**:

    - **IDE compatibility**: The **integrated development environment** (IDE) used for system development should support the programming language of an LLM. Most modern IDEs such as Visual Studio Code, PyCharm, and Eclipse offer support for multiple languages and tools for debugging and version control.

    - **Version control**: When integrating LLMs, it's important to use version control systems such as Git to manage changes to the code base. This allows teams to collaborate effectively, track changes, and revert to earlier versions if needed.

    - **Build and deployment tools**: Tools for building and deploying applications, such as Jenkins for **continuous integration / continuous deployment (CI/CD)** pipelines, should be compatible with the language and the LLM integration process.

The following are considerations for integration:

- **API integration**:

  - **RESTful APIs**: LLMs can also be accessed via RESTful APIs, which are language-agnostic. This means that regardless of the programming language used in the current system, the LLM can be accessed over HTTP.

  - **gRPC and other protocols**: For systems that require high-performance communication between services, protocols such as gRPC, which is supported by languages such as Go, Java, and C#, can be used.

- **Cross-language integration**:

  - **Interoperability**: If the system is built in a language different from an LLM's SDK, interoperability mechanisms such as **Foreign Function Interfaces** (**FFI**) or cross-language services such as Apache Thrift may be needed.

  - **Microservices architecture**: Adopting a microservices architecture can allow different services to be written in different languages, which is helpful if an LLM is best supported by a different language than the existing system.

- **Future-proofing**:

  - **Language trends**: Consider the longevity and support for the programming language. Languages that are widely adopted and supported are less likely to become obsolete and have a greater chance of being supported by future tools and technologies.

  - **Scalability**: Ensure that the programming language and the development practices adopted can scale with the growth of a system and the increasing complexity that comes with LLM integration.

- **Security**:

  - **Secure coding practices**: Regardless of the language, follow secure coding practices to protect the system and the LLM from vulnerabilities.

  - **Dependency management**: Regularly update libraries and dependencies to patch security vulnerabilities and maintain compatibility with the LLM.

In summary, the programming language of the current system and the LLM should be conducive to integration. Utilizing an LLM with a native SDK in the same language as the existing system simplifies the process. However, with the proper tools and strategies, it is possible to integrate LLMs across different programming languages. The key is to prioritize compatibility, community support, performance, and future scalability to ensure a successful integration.

APIs

The use of APIs is central to the integration of LLMs into existing systems. Here's a detailed look at the role of APIs in LLM integration and the best practices to ensure a seamless connection.

The following are relevant for understanding APIs for LLMs:

- **API types and functions**:

  - **RESTful APIs**: **Representational State Transfer** (**REST**) APIs are the most common API type used for web services. They use HTTP requests to perform GET, PUT, POST, and DELETE actions on data. RESTful APIs are stateless, meaning that each request from a client to a server must contain all the information needed to understand and complete the request.

  - **GraphQL APIs**: An alternative to RESTful APIs, GraphQL allows clients to request exactly the data they need, making it an efficient way to interact with LLMs, especially when dealing with large datasets.

  - **gRPC APIs**: gRPC is a modern, open source **remote procedure call** (**RPC**) framework that can run in any environment. It uses HTTP/2 for transport and Protocol Buffers as the interface description language, and it provides features such as authentication and load balancing.

- **API endpoints**:

  - An API endpoint functions as the gateway for data exchange in the interface between two interacting systems, marking where an API meets a server

The following factors are important to ensure API consumption:

- **Compatibility checks**:

  - Before integration, it's important to check that the existing system is capable of sending requests to and receiving responses from an LLM's API. This includes being able to handle the correct HTTP methods, headers, and data formats (such as JSON or XML).

- **API management policies**:

  - **Rate limiting**: Ensure that an LLM's API can handle the expected request load without violating rate limits. If the existing system's demands exceed the LLM's API limits, it might be necessary to implement a queuing system or look for an LLM provider that can meet higher request volumes.

  - **Authentication and authorization**: Verify that the API's security protocols are compatible with the existing system. This often involves the use of API keys, OAuth tokens, or other credentials that must be managed securely.

- **Error handling**: The existing system must be prepared to handle errors from the API gracefully. This includes proper logging of errors and the implementation of retry logic, where appropriate.

Here are the main best practices for API integration:

- **Documentation and testing**: Comprehensive documentation is crucial for understanding how to interact with an API effectively. Additionally, testing the API endpoints with tools such as Postman or automated scripts ensures that the integration works as expected before deployment.

- **Monitoring and maintenance**: Once integrated, an API's performance should be monitored using tools that can track response times, success rates, and error rates. Regular maintenance checks are also necessary to ensure that the API is updated as an LLM evolves and that any deprecated features are addressed.

- **Versioning**: API versioning is important to manage changes over time. When an LLM's API is updated, these versions ensure that the existing system can continue functioning with the current version while preparing to adapt to the new one.

- **Caching strategies**: Implement caching where appropriate to reduce the number of API calls and to improve the performance of a system. However, be mindful of the data freshness requirements, as LLMs often need the most up-to-date information to generate accurate outputs.

In summary, APIs are essential for integrating LLMs, as they define the methods and protocols for communication between an LLM and the existing system. Ensuring compatibility, adhering to API management policies, and following best practices for integration are all critical steps in achieving a successful and robust connection between systems. Properly managed, APIs can facilitate a smooth and efficient integration process, allowing organizations to leverage the powerful capabilities of LLMs in their existing technological ecosystems.

Frameworks

Frameworks in software development are foundational structures used to build and organize web applications, services, and other development projects. Next, we will take an in-depth look at the considerations for integrating LLMs with common frameworks.

Django for Python

Django is an advanced Python web framework that facilitates quick development and sensible, straightforward design. It aims to enable developers to progress applications from an initial idea to the final implementation rapidly.

Here's what you need to know about LLM integration with Django:

- **Django REST framework**: To integrate an LLM with a Django application, you can use the Django REST Framework to create API endpoints that interact with the LLM. This can involve sending data to the LLM for processing and retrieving the results to present to users, or further processing within the application.

- **Asynchronous tasks**: For LLMs that require longer processing times, you might need to use asynchronous task queues such as Celery with Django. This allows the Django application to send tasks to be processed in the background, without blocking the main application thread.

- **Middleware customization**: Django's middleware can be used to add functionality, such as automatically translating text or handling user input before it reaches the view or after the view has processed the request.

Spring for Java

In the context of contemporary Java-based enterprise applications, Spring offers a comprehensive programming and configuration framework that is adaptable to various deployment environments.

Here's what you need to know about LLM integration with Spring:

- **Spring Boot**: With Spring Boot, creating standalone, production-grade Spring-based applications that you can "just run" is straightforward. It simplifies the integration of LLMs by providing auto-configuration options and easy access to command-line interfaces for scripting LLM interactions.

- **Spring Cloud**: For cloud-based LLMs, Spring Cloud provides tools for developers to quickly build some of the common patterns in distributed systems (for example, configuration management and service discovery).

- **Spring Data REST**: This project makes it easy to build hypermedia-driven REST web services on top of Spring Data repositories. A Spring-based application can interact with an LLM through these REST services.

Middleware for adaptability

Middleware is software that facilitates communication and data management for distributed applications, positioned between an operating system and the applications running on it.

Here are some relevant tools for middleware integration:

- **Adaptors and connectors**: If an LLM is not natively compatible with a framework, middleware can act as a bridge. For example, a middleware adaptor can transform data from a Spring application into a suitable format for an LLM API and handle the response.

- **Enterprise Service Bus (ESB)**: An ESB can be used to integrate different applications by providing a communication bus between them. It can route data, transform it to the appropriate format, and handle various types of protocols and interfaces.

- **API gateways**: An API gateway functions as an intermediary reverse proxy positioned between a client and multiple backend services. It processes incoming API requests, orchestrates the required services to address these requests, and delivers the corresponding response. This tool manages the interactions.

The best practices for framework integration are as follows:

- **Documentation and support**: Leverage the extensive documentation and community support available for frameworks such as Django and Spring to implement best practices in LLM integration
- **Modularity**: Design the integration in a modular way so that changes in an LLM can be accommodated with minimal changes to an application
- **Security**: Ensure that the integration adheres to security best practices, particularly when the LLM is accessible over the web or involves sensitive data
- **Testing**: Implement comprehensive testing strategies, including unit tests, integration tests, and end-to-end tests, to ensure that the integration works as expected

Integrating an LLM into an existing system that relies on frameworks such as Django or Spring requires careful consideration of the framework's capabilities and constraints. By leveraging the tools and best practices provided by these frameworks, developers can create robust, scalable, and secure integrations that harness the power of LLMs within their applications.

Aligning with operational workflows

The integration of LLMs such as GPT-4 into operational workflows marks a significant leap forward in how businesses and individuals tackle various tasks. Let's review this further.

Process augmentation

Process augmentation through LLMs such as GPT-4 represents a significant advancement in how businesses and individuals approach various tasks.

Augmentation in customer service is concerned with the following:

- **Automated response systems**: LLMs can manage routine customer inquiries, providing instant responses to common questions. This reduces the response time and improves customer satisfaction. It also allows human customer service representatives to focus on more complex issues that require a human touch.
- **Personalization of interactions**: LLMs can analyze customer data to personalize interactions, ensuring that the responses are not just accurate but also tailored to the individual customer's history and preferences.

- **Feedback analysis**: LLMs can process and analyze customer feedback at scale, identifying common issues or trends that might require attention. This enables businesses to rapidly adapt to customer needs and improve their products or services.

Augmentation in marketing and content creation is concerned with the following:

- **Content generation**: LLMs can produce a wide range of content, from social media posts to blog articles. This can greatly assist in maintaining a consistent online presence, crucial for digital marketing strategies.

- **Idea generation and brainstorming**: Marketers can use LLMs to generate ideas for campaigns, slogans, or branding strategies. While the final creative decision-making remains in human hands, LLMs can provide a starting point or inspiration.

- **Language and tone adaptation**: LLMs can adapt content to different audiences by adjusting the language, tone, and style to suit various demographics or cultural backgrounds, ensuring broader appeal and effectiveness.

The key considerations for task selection are as follows:

- **The balance between human and machine input**: The primary goal should be to assist and enhance human work, not replace it. Tasks that require emotional intelligence, deep cultural understanding, or complex decision-making are often best handled by humans, with LLMs providing support.

- **Accuracy and reliability**: While LLMs are highly capable, they are not infallible. Businesses should assess the accuracy needs of a task and the risks associated with potential errors. For tasks where high accuracy is critical, human oversight is necessary.

- **Data privacy and ethical considerations**: When using LLMs, especially in customer service, it's crucial to consider data privacy and ethical implications. Businesses must ensure that the use of customer data complies with legal standards and respects customer privacy.

- **Continuous learning and adaptation**: LLMs can improve over time with feedback and additional training. Businesses should establish mechanisms for regular updates and training to keep the LLMs effective and relevant.

Future prospects

The use of LLMs in process augmentation is an area of rapid growth and innovation. As these models become more sophisticated, their potential applications will expand, offering more ways to augment human tasks effectively. However, the key to successful implementation will always lie in finding the right balance, ensuring that LLMs serve as a valuable tool that complements human skills and creativity, rather than attempting to replace them. This approach not only maximizes the benefits of LLMs but also safeguards the unique contributions that only humans can make.

Automation of tasks

LLMs such as GPT-4 are designed to handle a wide array of text-based tasks that are repetitive and follow specific patterns. The following is a detailed look at how LLMs can automate tasks:

- **Report generation:**

  - **Data-driven reports:** LLMs can generate reports by pulling information from structured data sources. They can be programmed to understand various data points and compile them into coherent and comprehensive reports.

  - **Customization and scalability:** Users can customize the parameters of the reports based on their needs. LLMs can scale this process, handling numerous reports simultaneously, which would be time-consuming for humans.

  - **Natural language explanations:** They can translate complex data into understandable narratives, making reports more accessible to stakeholders who may not have expertise in data analysis.

- **Document summarization:**

  - **Processing bulk information:** LLMs can quickly read and summarize long documents, identifying key points and themes without the need for human reading, which is beneficial for legal, academic, or corporate research

  - **Cross-document analysis:** When summarizing multiple documents, LLMs can identify and communicate the overarching narrative or trends across all texts

  - **Custom summaries:** Summaries can be tailored to the desired length and focus, whether executive summaries for leadership or abstracts for researchers

- **Email management:**

  - **Sorting and prioritization:** LLMs can manage and prioritize emails by urgency, topic, or sender, helping professionals focus on the most important communications first

  - **Automated responses:** For standard inquiries, LLMs can draft replies based on previous responses or templates, ensuring timely communication

- **Coding and scripting:**

  - **Boilerplate code generation:** For repetitive coding tasks, LLMs can generate boilerplate code, allowing developers to focus on more complex and creative aspects of software development

  - **Script automation:** They can write scripts for data analysis, file management, or system operations, automating routine technical tasks

- **Ensuring effective automation**:

  - **Integration with existing workflows**: For an automation to be successful, LLMs should be seamlessly integrated into existing workflows without disrupting them

  - **Training and fine-tuning**: LLMs may require initial training and fine-tuning to adapt to the specific requirements and context of the tasks they will automate

  - **Human oversight**: Despite their capabilities, LLMs should operate under human oversight to catch and correct errors, manage exceptions, and provide ethical judgment when necessary

The automation of repetitive and pattern-based tasks by LLMs has the potential to revolutionize many aspects of work, freeing up human resources for higher-level tasks that require creativity, critical thinking, and emotional intelligence. As these models continue to improve, their adoption will likely become more widespread, leading to further efficiency gains and the transformation of traditional job roles. However, striking a balance between automation and human oversight remains crucial to address the limitations of current AI technologies and ensure ethical and responsible use.

Customization needs

Customization of LLMs is a crucial step to ensure that they can effectively meet the unique needs of different businesses and industries. This process involves several key considerations:

- **Understanding industry-specific requirements**:

  - **Terminology**: Different industries have their own jargon and technical language. An LLM must understand and use this language correctly to be effective.

  - **Processes**: Each industry follows distinct processes, which an LLM should be able to navigate or reference accurately.

  - **Regulations**: Certain sectors are highly regulated, and LLMs must be configured to comply with relevant laws and guidelines.

- **Training on domain-specific datasets**:

  - **Dataset collection**: Gather text data that is representative of an industry's communication style, technical language, and common tasks.

  - **Dataset quality**: Ensure that the data is high-quality, relevant, and extensive enough to cover the scope of an LLM's intended applications.

  - **Model fine-tuning**: Use the collected datasets to fine-tune the LLM so that it better understands and generates industry-specific content.

- **Modifying outputs**:

  - **Tone and style**: An LLM's output should match the company's brand voice and communication style. This may require adjustments to the model's default generation style.

  - **Templates and formats**: For tasks such as report generation, the LLM should produce content that fits into a company's templates and formats.

- **Evaluation and iteration**:

  - **Feedback loops**: Establish mechanisms to gather feedback on an LLM's performance and use this feedback for continuous improvement

  - **Iterative training**: Regularly update the training datasets and fine-tune parameters to adapt to changes in industry language or company needs

- **Integration with existing systems**:

  - **APIs and interfaces**: Create interfaces for an LLM to interact with existing business systems, such as **customer relationship management (CRM)** or **enterprise resource planning (ERP)** systems

  - **Automation workflows**: Determine how the LLM will fit into current workflows and automate tasks without disruption

- **Considerations for optimal integration**:

  - **User training**: Ensure that the staff who will be working with an LLM are trained to understand its capabilities and limitations

  - **Performance monitoring**: Continuously monitor the LLM's performance to ensure that it meets the business objectives, and make adjustments as necessary

  - **Ethical and responsible use**: Maintain ethical standards, especially regarding data privacy and the avoidance of bias in model outputs

Customizing an LLM for a specific business or industry is a complex task that involves a deep understanding of the domain, careful preparation of training data, and ongoing monitoring and refinement of the model's performance. The customization must be approached as an iterative process, with the understanding that both the model and a company's needs will evolve over time. Properly customized LLMs can become powerful tools that enhance efficiency, improve customer experiences, and drive innovation within the company.

Outcome achievement

The integration of LLMs into business processes is not an end in itself but a means to achieve specific, valuable outcomes. To ensure that the deployment of an LLM leads to success, there are several key considerations to keep in mind.

- **Setting clear objectives**:

  - **Define success metrics**: Establish what success looks like for the integration of an LLM. This could be measured in terms of improved response time in customer service, higher engagement rates in content marketing, or more accurate data analysis.

  - **Align with business goals**: Ensure that the objectives for an LLM are in line with broader business goals. Whether it's to improve efficiency, reduce costs, or enhance the customer experience, the LLM's role should directly contribute to these targets.

  - **Specificity in objectives**: Be specific about what an LLM should accomplish. For example, rather than a vague goal of "*improving customer satisfaction*," aim for "*reducing average customer service response time by 50%.*"

- **Implementing and integrating an LLM**:

  - **Integration with systems**: Seamlessly integrate an LLM into existing systems without causing disruptions. This might require custom API development or the use of middleware.

  - **User experience**: Consider the end-user experience, whether it is the employees who interact with the LLM or the customers who receive its outputs.

  - **Iterative implementation**: Roll out the LLM in phases, starting with a pilot program to gauge effectiveness, and make necessary adjustments before full-scale implementation.

- **Monitoring and measuring performance**:

  - **Continuous monitoring**: Establish KPIs and use analytics to continuously monitor an LLM's performance against these indicators

  - **Feedback mechanisms**: Implement channels for users to provide feedback on the LLM's outputs and performance, contributing to continuous improvement

  - **Adaptation and optimization**: Use the performance data and feedback to refine the LLM's functioning, train it on additional data, or tweak its parameters to better achieve the set objectives

- **Evaluating impact**:

  - **Quantitative analysis**: Use statistical methods to measure the direct impact of an LLM on efficiency, productivity, and other quantifiable metrics

- **Qualitative assessment**: Evaluate the qualitative aspects, such as customer satisfaction or employee morale, which might be influenced by the LLM's integration

- **Cost-benefit analysis**: Consider the **return on investment (ROI)** by comparing the costs of implementing and maintaining the LLM against the financial and non-financial benefits gained

- **Ensuring scalability and sustainability**:

  - **Scalability**: Plan for scalability from the outset, ensuring that an LLM can handle increased loads or be expanded to additional tasks as needed

  - **Sustainability**: Ensure that the LLM's operations are sustainable, with mechanisms for regular updates, maintenance, and retraining to adapt to changing conditions

The ultimate goal of integrating LLMs into business operations is to achieve tangible, positive outcomes that align with strategic objectives. This involves careful planning, clear goal-setting, effective implementation, and ongoing management. The ability of an LLM to learn and adapt over time is one of its greatest strengths, and leveraging this capability can lead to continuous improvement in processes and outcomes. As LLM technology advances and becomes more sophisticated, its potential to transform businesses and contribute to achieving a wide range of outcomes will only increase.

To summarize, evaluating compatibility is about understanding the requirements and limitations of both an LLM and the current system, and then devising a strategy to bring them together in a manner that is technically sound and operationally harmonious. This process is critical not only for the successful deployment of LLMs but also for ensuring that their integration drives tangible value for an organization.

Seamless integration techniques

The seamless integration of LLMs into existing systems is a sophisticated endeavor that requires a strategic and methodical approach to minimize the impact on current operations. The goal is to ensure that LLMs enhance system functionality without causing significant disruption to existing workflows or user experiences. In the following subsections, each element of this multilayered strategy is broken down to elucidate the meticulous processes involved.

Incremental implementation

Incremental implementation in LLMs refers to gradually integrating and testing new features or improvements in the model, step by step, to enhance performance or capabilities over time.

Here are the details:

- **Phased rollout**: Instead of a "big bang" approach, a phased rollout allows for the gradual introduction of LLMs. This means starting with a pilot program or a specific department before expanding to the rest of an organization. It serves to reduce risk by allowing issues to be identified and resolved on a smaller scale before full deployment.

- **Learning and adaptation**: Both a system and its users will require time to adapt. For an LLM, this might mean initial training on smaller datasets and scaling up as the model's accuracy improves. For users, it might involve training sessions and the creation of new operational guidelines.

- **Feedback integration**: During the incremental implementation, user feedback is critical. This feedback should be used to adjust an LLM's integration, ensuring that it meets users' needs and fits into the existing workflows as smoothly as possible.

API and microservices architecture

In LLMs, an API facilitates the interaction between a language model and external applications. Microservices architecture involves structuring the model's features and components as a suite of small, independent services, improving scalability and maintenance.

Let's explore them further:

- **API-led connectivity**: APIs act as the connective tissue between LLMs and existing systems, allowing for the exchange of data and functionalities without extensive changes to the system's core. Well-defined APIs can simplify the process of updating and maintaining LLMs, as they provide a standardized way of accessing a model's capabilities.

- **Microservices for modularity**: Microservices architecture involves breaking down applications into smaller, loosely coupled services. By encapsulating LLM functionality in a microservice, it becomes easier to integrate, scale, and update without disrupting the entire system. This modularity also allows different parts of a system to evolve independently.

Data pipeline management

Data pipeline management for LLMs organizes and automates the collection, cleaning, and preparation of data used to train and update the models.

Let's explore this further:

- **Data quality assurance**: The performance of LLMs is highly dependent on the quality of data they are trained and run on. Ensuring that data is clean, well-structured, and representative is crucial. This might involve the use of data cleansing tools, ETL processes, and the development of schemas that define how data should be structured.

- **Pipeline reliability**: The data pipeline must be robust and capable of handling the volume and velocity of data required by LLMs. This involves considering data ingestion methods, storage solutions, and the orchestration of a data flow.

Monitoring and feedback loops

Monitoring and feedback loops in LLMs involve tracking the performance and behavior of the models in real time, and then using this data to continuously refine and improve their accuracy and efficiency. Let's explore this in more detail:

- **Performance metrics**: **Key performance indicators** (**KPIs**) must be established to assess an LLM's impact. This includes metrics for accuracy, response time, and user satisfaction. Monitoring tools can be used to track these KPIs in real time, allowing for proactive management of the LLM's performance.

- **Continuous improvement**: By setting up feedback loops, both from system monitoring and user input, organizations can implement a continuous improvement process. This ensures that an LLM remains effective and that its integration remains aligned with user requirements and system evolution.

- **Adaptive learning**: Some LLMs can improve over time through machine learning techniques. Implementing adaptive learning mechanisms, where an LLM can learn from interactions and improve its performance, can be part of the feedback loop.

In summary, effectively integrating LLMs involves a strategic approach that includes gradually implementing an LLM with APIs and microservices for modularity, managing data pipelines, and establishing robust monitoring. These steps ensure that the LLM fits seamlessly into existing systems and adapts to evolving organizational needs with minimal disruption.

Customizing LLMs for system-specific requirements

Customizing LLMs to meet system-specific requirements is pivotal to maximizing their efficiency and relevance in a given context. Tailoring these models allows them to perform optimally within the unique constraints and demands of different industries or business functions. Here is a detailed analysis of the customization process.

Fine-tuning

Fine-tuning for LLMs involves adjusting a pre-trained model on a specific dataset to tailor its responses to particular tasks or domains.

Let's explore this in more detail:

- **Dataset selection**: The process begins by selecting a dataset that closely mirrors the language, terminology, and style of the target domain. For instance, if an LLM is to be used in the medical field, the dataset should be rich in medical jargon and patient-doctor interactions.

- **Model training**: The selected dataset is then used to further train or fine-tune the LLM. This process adjusts the model's weights and biases to make it more adept at understanding and generating text that is specific to a domain.

- **Performance evaluation**: After fine-tuning, a model's performance is evaluated using domain-specific metrics. For example, in a legal application, the model might be evaluated on its ability to accurately generate contract clauses.

- **Iterative refinement**: Fine-tuning is often an iterative process, with multiple rounds of training and evaluation to continually refine a model until it meets the desired performance threshold.

Adding domain-specific knowledge

Adding domain-specific knowledge to LLMs involves incorporating specialized information or data into a model to enhance its expertise and performance in specific fields or industries.

Let's explore this further:

- **Knowledge integration**: This can involve methods such as incorporating additional training data from a domain into a model, or providing the model with access to external databases or knowledge bases that it can query for information

- **Knowledge base linking**: For instance, an LLM used by a financial institution might be linked to up-to-date databases containing market data, financial regulations, and economic indicators

- **Dynamic learning**: Some advanced LLMs can be designed to dynamically incorporate new information into their knowledge base, allowing them to stay up to date with the latest developments in their domain

User interface adaptation

User interface adaptation for LLMs involves customizing the way users interact with a language model, ensuring that the interface meets specific user needs and preferences.

Let's explore this in more detail:

- **User interface customization**: The **user interface** (**UI**) through which users interact with an LLM must be designed to be intuitive and tailored to the specific workflows of a system. For example, a content management system might feature a UI that allows marketers to easily generate and edit copy using the LLM.

- **Integration with existing tools**: Often, a UI will need to integrate seamlessly with existing tools and platforms already in use. This might involve the creation of plugins or extensions for software such as CRM systems or ERP software.

- **Accessibility and usability**: A UI should also be accessible to all users, regardless of their technical expertise, and should support the tasks they need to perform with minimal complexity.

- **Feedback mechanisms**: Incorporating feedback mechanisms within a UI can help to collect user responses to an LLM's output, which can be used for further model refinement.

Customizing LLMs demands an understanding of their technical capabilities and the specific requirements of their application domain. By tuning models with targeted data, embedding domain knowledge, and adjusting user interfaces to fit workflows, LLMs can be adapted to address the unique needs of any industry, enhancing their relevance and integration into existing systems.

Addressing security and privacy concerns in integration

The integration of LLMs within existing systems, while offering substantial benefits, also introduces a variety of security and privacy challenges. Let's look in depth at the strategies and considerations involved in each of the outlined components:

- **Data privacy**:

  - **Encryption**: Encryption serves as the first line of defense in protecting data. It is crucial for organizations to implement robust encryption standards such as **AES** (**Advanced Encryption Standard**) for data at rest and **TLS** (**Transport Layer Security**) for data in transit. Encryption keys should also be managed securely, using services such as **hardware security modules** (**HSMs**) or key management services that provide centralized control over cryptographic keys.

  - **Access control**: Access control mechanisms should be context-aware, granting permissions based on factors such as the user role, the location, the time of access, and the sensitivity of the data being accessed. This means implementing a dynamic access control policy that can evaluate the risk of a data request in real time, with permissions adjusted accordingly.

  - **Data masking**: Data masking, or obfuscation, should be dynamic, allowing for different views of the same data for different users. Dynamic data masking solutions can be integrated with existing databases and applications to provide real-time data transformation, based on user permissions.

- **Compliance with regulations**:

  - **Data anonymization**: Anonymization techniques must be irreversible to prevent the re-identification of individuals. Advanced techniques such as differential privacy can be employed to add noise to datasets, thereby providing a balance between data utility and privacy.

  - **Consent management**: Consent management should be a transparent process, with clear communication to users about what data is collected, how it is used, and the control they have over their data. This involves not just initial consent confirmation but also the provision of easy-to-use tools for users to view, modify, or revoke consent at any time.

- **Regular audits**: Audits should be both internal and external, with the latter performed by third-party organizations to ensure impartiality. Audits should assess both the technical aspects of data handling and the organizational processes in place to maintain compliance.

- **Security protocols**:

  - **Regular updates and patches**: A systematic approach to software maintenance is required, which includes automated systems to track, test, and deploy updates. Patch management tools can help streamline this process.

  - **Intrusion Detection Systems**: IDS should be complemented with a **security information and event management (SIEM)** system that aggregates and analyzes log data from across a network, providing a comprehensive view of the security posture and aiding in the detection of sophisticated attacks.

  - **Disaster recovery plans**: These plans should be detailed and regularly tested, with clear roles and responsibilities outlined for personnel during a recovery operation. The use of cloud services can also provide geographic redundancy and facilitate faster recovery times.

- **Bias and ethical considerations**:

  - **Bias detection**: Tools to detect bias should be integrated into the development and deployment pipeline of LLMs. These tools should be capable of both statistical analysis to detect patterns of bias and semantic analysis to understand the context of potential bias.

  - **Diverse training data**: The selection of training data should involve stakeholders from diverse backgrounds and should be guided by principles of representativeness and inclusivity. This may involve actively sourcing data from underrepresented groups to ensure that a model has a broad and fair understanding of different languages, dialects, and cultural contexts.

  - **Ethical guidelines**: These guidelines should be developed with multi-stakeholder input, including ethicists, domain experts, legal advisors, and potentially even representatives from the user base. They should be living documents, updated regularly to reflect new insights and societal norms.

  - **Impact assessments**: Impact assessments should not be one-off events but part of a continuous process aligned with a product life cycle. These assessments should feed into a governance framework that can make informed decisions about the deployment, scaling, and potential withdrawal of LLM functionalities.

In addressing these security and privacy concerns, organizations must adopt a proactive and holistic approach. This includes not only deploying technical measures but also fostering a culture of security and ethical awareness across all levels of an organization. Additionally, user education is critical, as informed users are better equipped to make decisions about their data and to understand the implications of interacting with LLMs. By building a secure and ethical foundation for the integration of LLMs, organizations can not only ensure compliance and security but also build lasting trust with their users and stakeholders, a trust that is vital for the sustainable and responsible use of AI technologies.

Summary

In this chapter, we outlined the multifaceted process of integrating LLMs into existing systems, emphasizing the need for a detailed assessment of technical specifications, such as computing power, storage, and data access speed, to ensure compatibility with current infrastructures. We discussed the importance of processor requirements, GPU acceleration, and distributed storage systems in handling the data-intensive operations of LLMs. We also went into the nuances of data formats and the necessity for transformation processes, utilizing tools such as ETL and APIs, to maintain efficient workflows.

Furthermore, we highlighted the role of programming languages, frameworks, and APIs in facilitating seamless integration and communication between LLMs and current systems, ensuring that any new infrastructure is scalable and future-proof. We emphasized the need for a balance between augmenting processes and automating tasks while customizing LLMs to meet industry-specific requirements, all the while prioritizing security and privacy to uphold the integrity and trustworthiness of AI technologies within operational ecosystems.

In the next chapter, we will introduce advanced optimization techniques for performance.

9

Optimization Techniques for Performance

Optimization is the heart of this chapter, where you will be introduced to advanced techniques that improve the performance of LLMs without sacrificing efficiency. We will explore advanced techniques, including quantization and pruning, along with approaches for knowledge distillation. A targeted case study on mobile deployment will offer practical perspectives on how to effectively apply these methods.

In this chapter, we're going to cover the following main topics:

- Quantization – doing more with less
- Pruning – trimming the fat from LLMs
- Knowledge distillation – transferring wisdom efficiently
- Case study – optimizing an LLM for mobile deployment

Upon completing this chapter, you will have acquired a detailed knowledge of sophisticated techniques that enhance LLM performance while ensuring efficiency.

Quantization – doing more with less

Quantization is a model optimization technique that converts the precision of the numbers used in a model from higher precision formats, such as 32-bit floating-point, to lower precision formats, such as 8-bit integers. The main goals of quantization are to reduce the model size and to make it run faster during inference, which is the process of making predictions using the model.

When quantizing an LLM, several key benefits and considerations come into play, which we will discuss next.

Model size reduction

Model size reduction via quantization is an essential technique for adapting LLMs to environments with limited storage and memory. The process involves several key aspects:

- **Bit precision**: Traditional LLMs often use 32-bit floating-point numbers to represent the weights in their neural networks. Quantization reduces these to lower-precision formats, such as 16-bit, 8-bit, or even fewer bits. The reduction in bit precision directly translates to a smaller model size because each weight consumes fewer bits of storage.

- **Storage efficiency**: By decreasing the number of bits per weight, quantization allows the model to be stored more efficiently. For example, an 8-bit quantized model will require one-fourth of the storage space of a 32-bit floating-point model for the weights alone.

- **Distribution**: A smaller model size is particularly advantageous when it comes to distributing a model across networks, such as downloading a model onto a mobile device or deploying it across a fleet of IoT devices. The reduced size leads to lower bandwidth consumption and faster download times.

- **Memory footprint**: During inference, a quantized model occupies less memory, which is beneficial for devices with limited RAM. This reduction in memory footprint allows more applications to run concurrently or leaves more system resources available for other processes.

- **Trade-offs**: The primary trade-off with quantization is the potential loss of model accuracy. As precision decreases, the model may not capture the same subtle distinctions as before. However, advanced techniques such as quantization-aware training can mitigate this by fine-tuning the model weights within the constraints of lower precision.

- **Hardware compatibility**: Certain specialized hardware, such as edge TPUs and other AI accelerators, are optimized for low-precision arithmetic, and quantized models can take advantage of these optimizations for faster computation.

- **Energy consumption**: Lower precision computations typically require less energy, which is crucial for battery-powered devices. Quantization, therefore, can extend the battery life of devices running inference tasks.

- **Implementation**: Quantization can be implemented post-training or during training. Post-training quantization is simpler but may lead to greater accuracy loss, whereas quantization-aware training incorporates quantization into the training process, usually resulting in better performance of the quantized model.

Inference speed

Inference speed is a critical factor in the deployment of neural network models, particularly in scenarios requiring real-time processing or on devices with limited computational resources. The inference phase is where a trained model makes predictions on new data, and the speed of this process can be greatly affected by the precision of the computations involved.

Let's explore this in further detail:

- **Hardware accelerators**: CPUs and GPUs are commonly used hardware accelerators that can process mathematical operations in parallel. These accelerators are optimized to handle operations at specific bitwidths efficiently. Bitwidth refers to the number of bits a processor, system, or digital device can process or transfer in parallel at once, determining its data handling capacity and overall performance. Many modern accelerators are capable of performing operations with lower-bitwidth numbers much faster than those with higher precision.

- **Reduced computational intensity**: Operations with lower precision, such as 8-bit integers instead of 32-bit floating-point numbers, are less computationally intensive. This is because they require less data to be moved around on the chip, and the actual mathematical operations can be executed more rapidly.

- **Optimized memory usage**: Lower precision also means that more data can fit into an accelerator's memory (such as cache), which can speed up computation because the data is more readily accessible for processing.

- **Real-time applications**: For applications such as voice assistants, translation services, or **augmented reality** (**AR**), inference needs to happen in real time or near-real time. Faster inference times make these applications feasible and responsive.

- **Resource-constrained devices**: Devices such as smartphones, tablets, and embedded systems often have constraints on power, memory, and processing capabilities. Optimizing inference speed is crucial to enable advanced neural network applications to run effectively on these devices.

- **Energy efficiency**: Faster inference also means that a task can be completed using less energy, which is particularly beneficial for battery-powered devices.

- **Quantization and inference**: Quantization can significantly contribute to faster inference speeds. By reducing the bitwidth of the numbers used in a neural network, quantized models can take advantage of the optimized pathways in hardware designed for lower precision, thereby speeding up the operations.

- **Batch processing**: Along with precision, the ability to process multiple inputs at once (batch processing) can also speed up inference. However, the optimal batch size can depend on the precision and the hardware used.

Power efficiency

Power efficiency is a vital consideration in the design and deployment of computational models, particularly for battery-operated devices such as mobile phones, tablets, and wearable tech. Here's how power efficiency is influenced by different factors:

- **Lower precision arithmetic**: Arithmetic operations at lower bitwidths, such as 8-bit or 16-bit calculations rather than the standard 32-bit or 64-bit, inherently consume less power. This is due to several factors, including a reduction in the number of transistors switched during each operation and the decreased data movement, both within the CPU/GPU and between the processor and memory.

- **Reduced energy consumption**: When a processor performs operations at a lower precision, it can execute more operations per energy unit consumed compared to operations at a higher precision. This is especially important for devices where energy conservation is crucial, such as mobile phones, where battery life is a limiting factor for user experience.

- **Thermal management**: Lower power consumption also means less heat generation. This is beneficial for a device's thermal management, as excessive heat can lead to throttling down the CPU/GPU speed, which in turn affects performance and can cause discomfort to the user.

- **Inference efficiency**: In the context of neural networks, most of the power consumption occurs during the inference phase when a model makes predictions. Lower precision during inference not only speeds up the process but also reduces power usage, allowing for more inferences per battery charge.

- **Voltage and current reductions**: Power consumption in digital circuits is related to the voltage and the current. Lower precision operations can often be performed with lower voltage and current levels, contributing to overall power efficiency.

- **Quantization benefits**: Since quantization reduces the precision of weights and activations in neural networks, it can lead to significant power savings. When combined with techniques such as quantization-aware training, it's possible to achieve models that are both power-efficient and maintain high levels of accuracy.

- **Optimized hardware**: Some hardware is specifically designed to be power-efficient with low-precision arithmetic. For example, edge TPUs and other dedicated AI chips often run low-precision operations more efficiently than general-purpose CPUs or GPUs.

- **Battery life extension**: For devices such as smartphones that are used throughout the day, power-efficient models can significantly extend battery life, enabling users to rely on AI-powered applications without frequently needing to recharge.

Hardware compatibility

Hardware compatibility is a critical aspect of deploying neural network models, including LLMs, particularly on edge devices. Edge devices such as mobile phones, IoT devices, and other consumer electronics often include specialized hardware accelerators that are designed to perform certain types of computations more efficiently than general-purpose CPUs. Let's take a deeper look into how quantization enhances hardware compatibility:

- **Specialized accelerators**: These are often **Application-Specific Integrated Circuits (ASICs)** or **field-programmable gate arrays (FPGAs)** optimized for specific types of operations. For AI and machine learning, many such accelerators are optimized for low-precision arithmetic, which allows them to perform operations faster, with less power, and more efficiently than high-precision arithmetic.

- **Quantization and accelerators**: Quantization adapts LLMs to leverage these accelerators by converting a model's weights and activations from high-precision formats (such as 32-bit floating-point) to lower-precision formats (such as 8-bit integers). This process ensures that models can utilize the full capabilities of these specialized hardware components.

- **Efficient execution**: By making LLMs compatible with hardware accelerators, quantization enables efficient execution of complex computational tasks. This is particularly important for tasks that involve processing large amounts of data or require real-time performance, such as natural language understanding, voice recognition, and on-device translation.

- **A wider range of hardware**: Quantization expands the range of hardware on which LLMs can run effectively. Without quantization, LLMs might only run on high-end devices with powerful CPUs or GPUs. Quantization allows these models to also run on less powerful devices, making the technology accessible to a broader user base.

- **Edge computing**: The ability to run LLMs on edge devices aligns with the growing trend of edge computing, where data processing is performed on the device itself rather than in a centralized data center. This has benefits for privacy, as sensitive data doesn't need to be transmitted over the internet, and for latency, as the processing happens locally.

- **Battery-powered devices**: Many devices are battery-powered and have strict energy consumption requirements. Hardware accelerators optimized for low-precision arithmetic can perform the necessary computations without draining the battery, making them ideal for mobile and portable devices.

- **AI at the edge**: With quantization, LLMs become a viable option for a wide range of applications that require AI at the edge. This includes not just consumer electronics but also industrial and medical devices, where local data processing is essential.

A minimal impact on accuracy

Quantization reduces the precision of a model's parameters from floating-point to lower-bitwidth representations, such as integers. This process can potentially impact the model's accuracy due to the reduced expressiveness of the parameters. However, with the following careful techniques, accuracy loss can be minimized:

- **Quantization-aware training**: This involves simulating the effects of quantization during the training process. By incorporating knowledge of the quantization into the training, a model learns to maintain performance despite the reduced precision. The training process includes the quantization operations within the computation graph, allowing the model to adapt to the quantization-induced noise and find robust parameter values that will work well when quantized.

- **Fine-tuning**: After the initial quantization, the model often undergoes a fine-tuning phase where it continues to learn with the quantized weights. This allows the model to adjust and optimize its parameters within the constraints of lower precision.

- **Precision selection**: Not all parts of a neural network may require the same level of precision. By selecting which layers or parts of a model to quantize, and to what degree, it's possible to balance performance with model size and speed. For example, the first and last layers of the network might be kept at higher precision, since they can disproportionately affect the final accuracy.

- **Calibration**: This involves adjusting the scale factors in quantization to minimize information loss. Proper calibration ensures that the dynamic range of the weights and activations matches the range provided by the quantized representation.

- **Hybrid approaches**: Sometimes, a hybrid approach is used where only certain parts of a model are quantized, or different precision levels are used for different parts of the model. For instance, weights might be quantized to 8-bit while activations are quantized to 16-bit.

- **Loss scaling**: During training, adjusting the scale of the loss function can help the optimizer focus on the most significant errors, which can be important when training with quantization.

- **Cross-layer equalization and bias correction**: These are techniques to adjust the scale of weights and biases across different layers to minimize the quantization error.

- **Data augmentation**: This helps a model generalize better and can indirectly help maintain accuracy after quantization by making the model less sensitive to small perturbations in the input data.

Trade-offs

Quantization of neural network models, including LLMs, brings significant benefits in terms of model size, computational speed, and power efficiency, but it is not without its trade-offs, such as the following:

- **Accuracy loss**: The primary trade-off with quantization is the potential for reduced model accuracy. High-precision calculations can capture subtle data patterns that might be lost when precision is reduced. This is particularly critical in tasks requiring fine-grained discrimination, such as distinguishing between similar language contexts or detecting small but significant variations in input data.

- **Model complexity**: Some neural network architectures are more sensitive to quantization than others. Complex models with many layers and parameters, or models that rely on precise calculations, may see a more pronounced drop in performance post-quantization. It may be harder to recover their original accuracy through fine-tuning or other optimization techniques.

- **Quantization granularity**: The level of quantization (that is, how many bits are used) can vary across different parts of a model. Choosing the right level for each layer or component involves a complex trade-off between performance and size. Coarse quantization (using fewer bits) can lead to greater efficiency gains but at the risk of higher accuracy loss, whereas fine quantization (using more bits) may retain more accuracy but with less benefit to size and speed.

- **Quantization-aware training**: To mitigate accuracy loss, quantization-aware training can be employed, which simulates the effects of quantization during the training process. However, this approach adds complexity and may require longer training times and more computational resources.

- **Expertise required**: Properly quantizing a model to balance the trade-offs between efficiency and accuracy often requires expert knowledge of neural network architecture and training techniques. It's not always straightforward and may involve iterative experimentation and tuning.

- **Hardware limitations**: The benefits of quantization are maximized when the target hardware supports efficient low-bitwidth arithmetic. If the deployment hardware does not have optimized pathways for quantized calculations, some of the efficiency gains may not be realized.

- **Model robustness**: Quantization can sometimes introduce brittleness in a model. The quantized model might not generalize as well to unseen data or might be more susceptible to adversarial attacks, where small perturbations to the input data cause incorrect model predictions.

- **Development time**: Finding the right balance between model size, accuracy, and speed often requires a significant investment in development time. The process can involve multiple rounds of quantization, evaluation, and adjustment before settling on the best approach.

Quantization is part of a broader set of model compression and optimization techniques aimed at making LLMs more practical for use in a wider array of environments, particularly those where computational resources are at a premium. It enables the deployment of sophisticated AI applications on everyday devices, bringing the power of LLMs into the hands of more users and expanding the potential use cases for this technology.

Pruning – trimming the fat from LLMs

Pruning is an optimization technique used to streamline LLMs by systematically removing parameters (that is, weights) that have little to no impact on the output. The main objective is to create a leaner model that retains essential functionality while being more efficient to run. Let's take a more detailed look at pruning.

The identification of redundant weights

The process of pruning a neural network, including LLMs, involves reducing the model's complexity by removing weights that are considered less important for the model's decision-making process. Here's a deeper insight into how redundant weights are identified and managed:

- **Weight magnitude**: Typically, the magnitude of a weight in a neural network indicates its importance. Smaller weights (closer to zero) have less impact on the output of the network. Therefore, weights with the smallest absolute values are often considered first for pruning.

- **Sensitivity analysis**: This involves analyzing how changes to weights affect a model's output. If the removal of certain weights does not significantly change the output or performance, these weights can be considered redundant.

- **Contribution to loss**: Weights can be evaluated based on their contribution to a model's loss function. Weights that contribute very little to reducing loss during training are candidates for removal.

- **Activation statistics**: Some pruning methods look at the activation statistics of neurons. If a neuron's output is frequently near zero, it's not contributing much to the next layer, and the weights leading into it might be pruned.

- **Regularization techniques**: L1 regularization promotes sparsity in the network weights. During training, L1 regularization can help identify weights that are less important, as they tend toward zero.

- **Pruning criteria**: Different pruning methods use different criteria to select weights to prune, such as gradient-based, Hessian-based, or Taylor expansion-based criteria, which consider the effect of the weight on model output more holistically. Other pruning criteria include dynamic pruning, magnitude pruning, gradient-based pruning, and group lasso pruning.

- **Global versus layer-wise pruning**: Pruning can be performed on a per-layer basis, where weights are pruned independently in each layer, or globally across the entire network. Global pruning considers the smallest weights across a whole network rather than within each layer.

- **Iterative pruning**: A network is often pruned iteratively, where a small percentage of weights are pruned at each iteration, followed by a period of retraining. This gradual process allows a network to adapt and compensate for the lost weights.

- **Pruning schedules**: These define when and how much pruning occurs during the training process. A schedule can be based on the number of epochs, a set performance threshold, or other training dynamics.

- **Validation**: After pruning, it's crucial to validate the pruned model on a held-out dataset to ensure that performance remains acceptable and that no critical weights have been removed.

Weight removal

In the context of optimizing neural networks, including LLMs, weight removal through pruning is a critical step following the identification of weights that contribute minimally to a network's output. Here's a detailed look into the process and implications of weight removal:

- **Pruning by zeroing weights**: The act of "pruning" refers to setting the identified less important weights to zero. It's akin to cutting off branches from a tree – the branch is no longer active or bearing fruit, although it remains part of the tree. Similarly, zeroed weights remain part of the network architecture but do not contribute to the calculations during forward and backward propagation.

- **Sparse network**: The result of pruning is a sparser network, where a significant number of weights are zero. Sparsity in this context means that there is a high proportion of zero-value weights relative to non-zero weights within the matrix that represents the network's parameters.

- **Maintained architecture size**: Even though many weights are set to zero, the overall architecture of a network does not change. The number of layers and the number of neurons within each layer remain the same, which means the metadata describing the network structure does not need to be altered.

- **Storage format**: Although a pruned network has the same dimensional architecture, it can be stored more efficiently if a sparse matrix format is used. Sparse formats store only non-zero elements and their indices, which can significantly reduce the storage space required for the network.

- **Computational efficiency**: While a network structure's size in terms of architecture remains the same, the actual number of computations required during inference is reduced. This is because multiplications by zero can be skipped, leading to faster processing times, especially if the hardware or software used for inference is optimized for sparse computations.

- **Implications for inference**: In practice, the computational benefits during inference depend on the level of support for sparse operations in the hardware and software. Some specialized hardware accelerators can take advantage of sparsity for increased efficiency, while others may not, resulting in no real speed-up.

- **Fine-tuning post-pruning**: After pruning, networks often undergo a fine-tuning process. This allows remaining non-zero weights to adjust and compensate for the loss of pruned weights, which can help recover any lost accuracy or performance.

- **Impact on overfitting**: Interestingly, pruning can sometimes improve the generalization of a network by removing weights that may contribute to overfitting on the training data. This can lead to improved performance on unseen test data.

- **Recovery of performance**: Pruning is typically an iterative process where a small percentage of weights are pruned at a time, followed by a period of retraining. This allows a network to maintain or even improve its performance despite the reduction in the number of active weights.

Sparsity

Sparsity in neural networks, such as LLMs, is a concept that arises from pruning, where certain weights within a network are set to zero. This results in a model that has a significant number of weights that do not contribute to the signal propagation in the network. Here are some important points about sparsity:

- **Sparse matrix**: In the context of neural networks, a sparse matrix is one where most of the elements are zero. This is in contrast to a dense matrix, where most elements are non-zero. Sparsity is a direct consequence of the pruning process.

- **Proportion of zero-valued weights**: Sparsity is quantitatively measured by the ratio of zero-valued weights to the total number of weights. A network is considered highly sparse if the majority of its weights are zero. For example, if 80% of the weights are zero, the network has 80% sparsity.

The benefits of sparsity include the following:

- **Memory efficiency**: Sparse models require less memory for storage, as the zero-valued weights can be omitted when using specialized sparse data structures

- **Computational efficiency**: During inference, calculations involving zero-valued weights can be skipped, potentially speeding up the process

- **Energy consumption**: Sparse operations typically consume less energy, which is beneficial for battery-powered devices

However, there are also some challenges with sparsity:

- **Hardware support**: Not all hardware is optimized for sparse computations. Some CPUs and GPUs are optimized for dense matrix operations and may not benefit from sparsity.

- **Software support**: Similarly, to leverage sparsity, the software performing the computations must be designed to handle sparse matrices efficiently.

The recommendations for the implementation of sparsity are as follows:

- **Sparse data structures**: To store sparse matrices efficiently, data structures such as **Compressed Sparse Row** (**CSR**) or **Compressed Sparse Column** (**CSC**) are used, which only store non-zero elements and their indices

- **Sparse operations**: Libraries and frameworks that support sparse operations can perform matrix multiplications and other calculations without processing the zero-valued elements

While high sparsity can make a model leaner and potentially faster, it can also lead to a decrease in model accuracy if too many informative weights are pruned.

Achieving high sparsity without significant loss of accuracy often requires careful iterative pruning and fine-tuning.

In practice, achieving sparsity in LLMs can be beneficial when deploying models to environments where resources are constrained, such as mobile phones, IoT devices, or edge servers.

Efficiency

In ML and neural network optimization, the term "efficiency" often refers to the ability to perform computations quickly and with minimal resource utilization. In the context of sparse models, efficiency gains are achieved through the structure of a neural network that has been pruned to contain many zero-valued weights. Here are the key points that contribute to the efficiency of sparse models:

- **Fewer computations**: Since the zero-valued weights do not contribute to the output, they do not need to be included in the computations. This means that the number of multiplications and additions during the forward and backward pass can be greatly reduced.

- **Optimized hardware**: There is specialized hardware that is designed to handle sparse matrix operations more efficiently than general-purpose processors. These can exploit the sparsity of a model to skip over zero-valued weights and only perform computations on the non-zero elements.

- **Quicker inference times**: With fewer computations required, a sparse model can produce outputs faster. This is crucial for applications that require real-time processing, such as natural language processing tasks, image recognition, or autonomous vehicle control systems.

- **Reduced memory usage**: Storing a sparse model requires less memory, since the zero-valued weights can be omitted. When using appropriate sparse matrix representations, only non-zero elements and their indices need to be stored. This can significantly reduce a model's memory footprint.

- **Bandwidth savings**: Transmitting a sparse model over a network requires less bandwidth than a dense model. This is beneficial when models need to be downloaded onto devices or updated frequently.

- **Energy conservation**: Sparse computations generally consume less energy, as many processing units can remain idle during operations. This makes sparse models particularly suitable for deployment on battery-operated devices, where energy efficiency is a priority.

- **Scalability**: Sparse models can be scaled to larger datasets and more complex problems without a proportional increase in computational resources. This scalability is beneficial for deploying advanced AI models on a wide range of hardware, from high-end servers to consumer-grade electronics.

- **Software support**: The efficiency of sparse models is also dependent on the software and libraries used to run them. Libraries that are optimized for sparse operations can efficiently execute a model's computations and fully utilize the hardware's capabilities.

The impact on performance

Pruning neural networks, such as LLMs, involves selectively removing weights, or connections, within a model that are deemed less important. The intent of pruning is to create a more efficient model without significantly compromising its accuracy or performance. A detailed examination of how pruning impacts performance is as follows:

- **Performance metrics**: A model's performance post-pruning is evaluated using various metrics, such as accuracy, precision, recall, and an F1 score for classification tasks. For LLMs involved in language tasks, perplexity, and a BLEU score might be used. These metrics assess how well the pruned model compares to its original version.

- **Iterative approach**: To mitigate the risk of performance loss, pruning is often performed iteratively. This means a small percentage of weights are removed at a time, and a model's performance is evaluated after each pruning step. If the performance metrics remain stable, further pruning can be considered.

- **Fine-tuning**: After each pruning iteration, a model is typically fine-tuned. This process involves additional training, allowing the model to adjust and optimize its remaining weights to recover from any accuracy loss due to pruning.

- **Aggressive pruning risks**: If pruning is too aggressive, a model might lose weights that are important for making accurate predictions, leading to a decrease in performance. This underscores the need for a cautious approach, where the pruning rate is carefully controlled.

- **Recovery of performance**: In some cases, a pruned model may even outperform the original model. This can occur because pruning helps to reduce overfitting by eliminating unnecessary weights, thereby improving the model's ability to generalize to new data.

- **Layer sensitivity**: Different layers in a neural network may have varying sensitivities to pruning. Pruning too much from a sensitive layer could result in a substantial performance drop, while other layers might tolerate more aggressive weight removal.

- **Hyperparameter tuning**: Post-pruning, hyperparameters of a model may need to be retuned. Learning rates, batch sizes, and other training parameters may require adjustment to accommodate the sparser structure of the model.

- **Resource-performance trade-off**: The impact on performance must be weighed against the benefits gained in efficiency. For deployment on resource-constrained devices, some loss in performance might be acceptable in exchange for gains in speed and reduction in model size.

- **Task-specific impact**: The acceptable degree of pruning can also depend on the specific task that an LLM is designed for. Tasks that rely on a nuanced understanding of language might suffer more from aggressive pruning than tasks that can tolerate some loss in detail.

Structured versus unstructured pruning

In the domain of neural network optimization, pruning is a common strategy used to reduce the size and computational complexity of models, including LLMs. There are two main types of pruning:

- **Unstructured pruning**:

  - This involves setting individual, specific weights within a network's weight matrix to zero

  - It creates a sparse matrix, where many weights are zero, but does not change the overall architecture of a model

  - The resulting model can still require the same computational resources if the hardware or software does not specifically optimize for sparse computations

  - Unstructured pruning is often easier to implement and can be done at a fine granularity, allowing for precise control over which weights are pruned

- **Structured pruning**:

  - Structured pruning removes entire neurons or filters (in the case of convolutional networks) rather than individual weights

  - This method can significantly reduce the complexity of a model because it removes entire sets of weights, thus simplifying the network architecture itself

  - Structured pruning can lead to models that are inherently smaller and may run faster on all types of hardware, not just those optimized for sparse computations

  - However, it can have a more pronounced impact on a model's performance, since it removes more of the model's capacity to represent and separate the data features

Both pruning techniques have their advantages and trade-offs:

- **Unstructured pruning**:

  - **Pros**: Allows you to fine-tune the pruning process and may retain more of a model's performance

  - **Cons**: May not reduce actual computational load unless specific sparse computation optimizations are in place

- **Structured pruning**:

  - **Pros**: Can lead to actual reductions in memory footprint and computational cost, regardless of hardware optimizations for sparsity

  - **Cons**: More likely to impact a model's performance due to the more significant reduction in model capacity

Pruning schedules

Pruning schedules are a strategic component of the model pruning process, particularly in the context of neural networks and LLMs. They are designed to manage the pruning process over time, with the goal of minimizing the negative impact on a model's performance. Here's a detailed exploration of pruning schedules:

- **Incremental pruning**: Instead of removing a large number of weights at once, pruning schedules typically involve incrementally pruning a small percentage of weights. This can occur after every epoch or after a predetermined number of epochs.

- **Compensation and adjustment**: By gradually pruning a model, the remaining weights have the opportunity to adjust during the retraining phases. This retraining allows a network to compensate for the lost connections and can lead to recovery of any lost accuracy or performance.

- **Phases of pruning and retraining**: A common approach in pruning schedules is to alternate between pruning and retraining phases. After each pruning phase, a network undergoes a period of retraining to fine-tune the remaining weights before the next round of pruning.

- **Determining pruning rate**: The schedule must define the rate at which weights are pruned. This rate can be constant or change over time. Some schedules may start with aggressive pruning rates that decrease over time as a model becomes more refined.

- **Criteria for pruning**: The schedule may also include criteria for selecting which weights to prune. This could be based on the magnitude of weights, their contribution to output variance, or other sophisticated criteria.

- **End criteria**: The schedule should specify an end criterion for pruning. This could be a target model size, a desired level of sparsity, a minimum acceptable performance metric, or simply a fixed number of pruning iterations.

- **Monitoring model performance**: Throughout the pruning process, it is crucial to continuously monitor a model's performance on a validation set. If performance drops below an acceptable threshold, the pruning schedule may need to be adjusted.

- **Pruning to threshold**: Some schedules prune based on a threshold value; weights below this threshold are pruned. This threshold can be adjusted throughout training to control the degree of pruning.

- **Automated stopping conditions**: Advanced pruning schedules may include automated stopping conditions that halt pruning if a model's performance degrades beyond a certain point.

- **Hyperparameter optimization**: Along with pruning, other hyperparameters of a network may need adjustment. Learning rates, for example, might be reduced after certain pruning thresholds are reached to stabilize training.

Fine-tuning

Fine-tuning is a crucial step in the model optimization process, particularly after pruning, which is the selective removal of weights in a neural network. Let's take an in-depth look at the fine-tuning process post-pruning:

- **The objective of fine-tuning**: The main goal of fine-tuning is to allow a model to adapt to the changes in its architecture that occurred due to pruning. Since pruning can disrupt the learned patterns within a network, fine-tuning aims to restore or even improve the model's performance by re-optimizing the remaining weights.

- **Training on a subset of data**: Fine-tuning does not typically require retraining from scratch on an entire dataset. Instead, it can be done on a subset or using fewer epochs, as the model has already learned the general features and only needs to adjust to the reduced complexity.

- **Learning rate adjustments**: During fine-tuning, the learning rate is often lower than during the initial training phase. This helps in making smaller, more precise updates to the weights, avoiding drastic changes that could destabilize a newly pruned model.

- **Recovering performance**: After pruning, there might be an initial drop in accuracy or an increase in loss. Fine-tuning helps to recover this lost performance by refining the weight values of the remaining connections, which compensates for the pruned ones.

- **Recalibration**: The process allows a model to recalibrate the importance of the remaining weights. It's possible that the dynamics of the network change after pruning, and fine-tuning helps a network find new paths for signal propagation, possibly leading to new and sometimes more efficient representations.

- **Iterative process**: In some cases, pruning and fine-tuning are done iteratively in cycles – pruning a bit, then fine-tuning, and then pruning again. This cyclic process can lead to a more gradual reduction in model size while maintaining performance.

- **Stochastic Gradient Descent (SGD)**: Fine-tuning is usually carried out using SGD or one of its variants, such as Adam or RMSprop. These optimizers are adept at finding good values for the weights, even in a highly pruned network.

- **Regularization techniques**: Techniques such as dropout or weight decay might be adjusted during fine-tuning to prevent overfitting, as the model capacity has been reduced due to pruning.

- **Performance monitoring**: It's essential to monitor performance closely during fine-tuning to ensure that a model is improving and not overfitting or diverging.

- **Stopping criteria**: Fine-tuning should have a clear stopping criterion based on performance metrics on a validation set, such as reaching a specific accuracy level or no longer seeing improvement over several epochs.

Pruning is an essential part of the model optimization toolkit, especially when deploying LLMs in environments with stringent computational or storage limitations. By reducing the computational load without substantial loss in output quality, pruning makes it feasible to utilize advanced neural networks in a wider range of applications and devices.

Knowledge distillation – transferring wisdom efficiently

Knowledge distillation is an effective technique for model compression and optimization, particularly useful for deploying sophisticated models such as LLMs on devices with limited resources. The process involves the aspects covered next.

Teacher-student model paradigm

Let's take a deeper dive into the concept of the teacher-student model paradigm in knowledge distillation:

- **Teacher model**: The "teacher" model serves as the source of knowledge in knowledge distillation. It is a well-established and usually complex neural network that has been extensively trained on a large dataset. This model has achieved high accuracy and is considered an expert in the task it was trained for. The teacher model serves as a reference or a benchmark for high-quality predictions.

- **Student model**: In contrast, the "student" model is a compact and simplified neural network with fewer parameters and layers compared to the teacher model. The purpose of the student model is to learn from the teacher model and replicate its behavior. Despite its reduced complexity, the student model aims to achieve comparable or close-to-comparable performance with the teacher model. Once the student model is trained, it can perform inference much faster and with lower memory requirements compared to the teacher model, with only a small sacrifice in accuracy. This makes the student model suitable for deployment in resource-constrained environments, such as mobile devices, embedded systems, or web applications.

- **Knowledge transfer**: Knowledge distillation is essentially a process of transferring the knowledge or expertise of the teacher model to the student model. This knowledge encompasses not only the final predictions but also the rich internal representations and insights that the teacher model has learned during its training.

- **Output mimicking**: The primary objective of the student model is to mimic the output probabilities of the teacher model. This means that when given an input, the student model should produce predictions that are similar to those of the teacher model. This output mimicking can be achieved through various techniques, including adjusting the loss function to penalize differences in predictions.

- **Loss function modification**: To facilitate knowledge transfer, the loss function during training is often modified. In addition to typical loss components such as cross-entropy, a distillation loss term is introduced. This term encourages the student model to match the soft targets (probability distributions) produced by the teacher model, rather than the hard targets (one-hot-encoded labels).

The benefits of knowledge distillation include the following:

- **Model compression**: Knowledge distillation results in a significantly smaller student model compared to the teacher model, making it suitable for deployment on resource-constrained devices such as mobile phones or edge devices

- **Improved efficiency**: The student model can make predictions faster than the teacher model due to its reduced complexity, which is valuable for real-time applications

- **Transferability**: Knowledge distillation can transfer knowledge across different model architectures and even across different tasks, enabling the student model to perform well in diverse scenarios

While knowledge distillation is a powerful technique, it's not without challenges. Finding the right balance between model complexity and performance, selecting suitable hyperparameters, and ensuring that the student model generalizes well can be non-trivial tasks.

The transfer of knowledge

The core objective of knowledge distillation is to transfer the "knowledge" acquired by the teacher model to the student model. This knowledge includes not only the final predictions made by the teacher model but also the rich insights and representations it has learned during its training on a large dataset.

This involves the following:

- **Teacher-student mismatch**: It's important to note that the teacher and student models can have different architectures. In fact, they often do. The teacher model is typically a larger, more complex neural network, while the student model is deliberately designed to be smaller and simpler. This architectural difference means that a straightforward parameter copy is not possible.

- **Emulating output distributions**: Instead of copying parameters, the student model is trained to emulate or replicate the output distributions generated by the teacher model. These output distributions can include class probabilities in classification tasks or any other relevant probability distributions for different types of tasks.

- **Loss function modification**: To achieve this emulation, the loss function used during training is modified. In addition to standard loss components such as cross-entropy, a distillation loss term is introduced. This distillation loss encourages the student model to produce output distributions that are as close as possible to those of the teacher model.

- **Soft targets versus hard targets**: In the context of knowledge distillation, the teacher model's predictions are often referred to as "soft targets" because they represent probability distributions over classes. In contrast, the traditional ground-truth labels used for training are "hard targets" because they are one-hot encoded. During training, the student model is provided with the "soft targets" from the teacher model. These soft targets are the output probabilities for each class, which carry more information than the "hard targets" of the true labels (which are just zeros and ones). For example, instead of just knowing that a particular image is of a "cat" (hard target), the student learns the degree of certainty (expressed in probabilities) that the teacher model attributes to that prediction (soft target).

- **Temperature parameter**: Another important aspect is the introduction of a temperature parameter in the distillation loss. This parameter controls the "softness" of the targets. A higher temperature leads to softer targets, which are more informative for training the student model. Conversely, a lower temperature results in harder targets that are closer to one-hot-encoded labels.

- **The benefits of output emulation**: Emulating the output distributions rather than directly copying parameters has several advantages. It allows the student model to capture the nuanced decision boundaries and uncertainty information present in the teacher model's predictions. This can lead to better generalization and more robust performance.

- **Practical applications**: Knowledge distillation is widely used in scenarios where model size and inference speed are critical, such as deploying models on mobile devices, edge devices, or in real-time applications. It allows you to create compact yet accurate models that are well-suited for resource-constrained environments.

Knowledge distillation trains a smaller student model to mimic the output distributions of a larger teacher model, enabling efficient and accurate inference in applications with limited computational resources. This technique is useful across fields such as language processing, computer vision, and speech recognition, particularly for deploying LLMs in resource-constrained environments.

Case study – optimizing the ExpressText LLM for mobile deployment

In this section, let's go through a hypothetical case study that exemplifies the optimization of an LLM for mobile deployment.

Background

ExpressText is a state-of-the-art LLM designed for NLP tasks, including translation and summarization. Despite its effectiveness, the model's size and computational demands limit its deployment on mobile devices.

Objective

The objective was to optimize ExpressText for mobile deployment, ensuring that it retains high accuracy while achieving a smaller size and faster inference on mobile hardware.

Methodology

Three main optimization techniques were applied:

- **Quantization**: The model's 32-bit floating-point weights were converted to 8-bit integers, significantly reducing its size. Quantization-aware training was employed to minimize accuracy loss.

- **Pruning**: Using iterative magnitude-based pruning, weights with the smallest absolute value were set to zero to create a sparser network. The model was pruned by 40% without substantial performance degradation.

- **Knowledge distillation**: A smaller "student" model was trained to mimic the "teacher" ExpressText's output distributions. Soft targets from the teacher and temperature scaling were used to transfer nuanced knowledge to the student.

Results

The optimized model achieved the following results:

- The model size was reduced from 1.5 GB to 300 MB, a five-fold decrease

- Inference speed improved by three times on standard mobile hardware

- 97% of the original model's accuracy was retained on benchmark tests

Challenges

The following challenges were faced:

- Balancing model size and accuracy, especially after aggressive pruning

- Ensuring that the student model captured nuanced language features from the teacher

- Adapting the quantization process to the model without significant latency issues

Solutions

To overcome the challenges, these solutions were implemented:

- A custom pruning schedule was developed to iteratively prune and fine-tune the model

- Extensive hyperparameter tuning was conducted during knowledge distillation to maintain performance

- Hardware-specific optimizations were implemented for different mobile platforms

Conclusion

The case study demonstrated that through careful application of quantization, pruning, and knowledge distillation, the ExpressText LLM could be effectively optimized for mobile deployment. The model maintained high accuracy while achieving a size and speed conducive to mobile environments, enabling its use in real-time language processing applications on smartphones and tablets.

This case study serves as an illustrative example of how optimization techniques can be applied to prepare complex LLMs for mobile deployment, addressing the constraints and requirements of mobile devices while preserving the functionality of a model.

Summary

In this chapter on performance optimization for LLMs, advanced techniques were introduced to enhance efficiency without compromising effectiveness. It discussed several methods, starting with quantization, which compresses models by reducing bit precision, thus shrinking model size and accelerating inference – a crucial phase where a model generates predictions. This involves a trade-off between model size and speed against accuracy, with tools such as quantization-aware training used to balance these aspects.

Pruning was another method discussed, focusing on eliminating less critical weights from LLMs to make them leaner and faster, which is particularly beneficial for devices with limited processing capabilities. Knowledge distillation was also covered, which involves transferring insights from a large, complex model (teacher) to a smaller, simpler one (student), retaining performance while ensuring that the model is lightweight enough for real-time applications or deployment on mobile devices.

The chapter concluded with a case study on mobile deployment, providing practical insights into how these optimization techniques can be implemented.

In the next chapter, we will continue exploring this topic, going further into advanced optimization and efficiency.

10

Advanced Optimization and Efficiency

Building on the previous chapter, we will dive deeper into the technical aspects of enhancing LLM performance. You will explore state-of-the-art hardware acceleration, and you will also learn how to manage data storage and representation for optimal efficiency and speed up inference without loss of quality. We will provide a balanced view of the trade-offs between cost and performance, a key consideration when deploying LLMs at scale.

In this chapter, we're going to cover the following main topics:

- Advanced hardware acceleration techniques
- Efficient data representation and storage
- Speeding up inference without compromising quality
- Balancing cost and performance in LLM deployment

By the end of this chapter, you will have acquired a comprehensive understanding of the technical intricacies involved in enhancing LLM performance beyond what was covered in the previous chapter.

Advanced hardware acceleration techniques

Advanced hardware acceleration techniques are pivotal in enhancing the capabilities of LLMs, by significantly boosting the speed and efficiency of necessary computations for their training and inference phases. Beyond the primary use of GPUs, TPUs, and FPGAs, let's explore some more sophisticated aspects and emerging trends in hardware acceleration that are pushing the boundaries of what's possible with LLMs.

Tensor cores

Tensor cores are a breakthrough in GPU architecture, designed to accelerate the matrix multiplications that power deep learning workloads. They enable mixed-precision arithmetic, a technique that uses different numerical precisions within the same computation. Here's how they contribute to deep learning:

- **Efficient matrix operations**: Tensor cores are optimized to perform the matrix multiplication and accumulation operations at the heart of neural network training and inference. They can carry out these operations in a fraction of the time it would take using traditional floating-point units.

- **Mixed-precision arithmetic**: The mixed-precision approach allows tensor cores to use lower-precision formats such as FP16 for the bulk of computations, while using higher-precision formats such as FP32 to accumulate results, striking a balance between speed and accuracy.

- **Boosted throughput**: With tensor cores, GPUs can deliver significantly higher throughput for deep learning operations, translating to faster model training and inference times.

Memory hierarchy optimization

Modern GPUs are designed with a complex memory hierarchy to address the following data movement challenges:

- **Shared memory**: A low-latency memory accessible by all threads in a block, which can be used to share data between threads and reduce global memory accesses.

- **Cache memory**: L1 and L2 caches in GPUs help to store frequently accessed data close to the compute cores, minimizing the need to access slower global memory.

- **Global memory**: The main memory pool from which data is loaded into caches and shared memory. Optimizing its usage is crucial, as global memory bandwidth can often be a limiting factor in GPU performance.

- **Memory bandwidth**: Advanced GPUs also feature high memory bandwidth, which is the rate at which data can be read from or stored in a semiconductor memory by a processor. Enhancements in memory technology such as **Graphics Double Data Rate 6** (**GDDR6**) and **High Bandwidth Memory** (**HBM2**) contribute to wider memory buses and higher data transfer speeds.

Asynchronous execution

Asynchronous execution in GPUs allows for better utilization of resources by supporting the following:

- **Concurrent kernel execution**: Modern GPUs can execute multiple kernels (the basic units of executable code that run on the GPU) concurrently, which can be particularly beneficial when those kernels don't fully utilize the GPU's resources.

- **Overlap of data transfer and computation**: While one kernel is running, data for the next can be transferred over the PCIe bus, thus overlapping computation with communication.

- **Stream multiprocessors**: Advanced GPUs contain multiple **stream multiprocessors (SMs)** that can handle different execution tasks simultaneously. Each SM can manage its own queue of operations, allowing multiple operations to be in flight at any given time.

- **Non-blocking algorithms**: Algorithms can be designed to be non-blocking, where tasks are divided into smaller chunks that can be processed independently, allowing other tasks to be performed in the gaps.

The integration of these advanced features results in GPUs that are not just faster but also smarter in how they manage computations and data. This is crucial for deep learning, where the ability to process large volumes of data quickly can be the difference between a feasible solution and an impractical one. For developers and researchers, leveraging these GPU features means they can train more complex models, experiment more rapidly, and deploy more sophisticated AI systems.

FPGAs' versatility and adaptability

Field-programmable gate arrays (FPGAs) are highly versatile and adaptable computing devices that are particularly useful in fields where the requirements can change over time, such as in the deployment of LLMs. Here's a closer look at the unique attributes of FPGAs:

- **Dynamic reconfiguration**:

  - **On-the-fly adaptability**: FPGAs are unique in their ability to be reconfigured while in use. This means that hardware can be programmed to perform different functions at different times, allowing a single FPGA to handle a variety of tasks that may be required at various stages of LLM processing.

  - **Rapid prototyping and testing**: Since FPGAs can be reprogrammed without the need for physical modifications, they are ideal for developing and testing new types of algorithms or model architectures. This can accelerate the prototyping phase of LLM development.

  - **Adaptive data processing**: As LLMs evolve, the FPGA can be reconfigured to support new models or updated algorithms, providing a level of future-proofing and ensuring that hardware remains relevant as the models become more advanced.

- **Precision tuning**:

  - **Customizable bitwidths**: FPGAs allow for the customization of precision down to the bit level. For LLMs, this means that a model can use exactly the precision it needs for different operations, which can optimize both the speed and the efficiency of the computations.

  - **Balancing accuracy and performance**: By adjusting the precision of arithmetic operations, FPGAs can find an optimal balance between the computational intensity of a task and the accuracy of the results. For example, an LLM might use lower precision for certain layers or operations where high precision is not critical, thereby saving resources and time.

- **Energy efficiency**: Lower precision calculations typically require less power, making FPGAs an energy-efficient option for running LLMs, especially in environments where power consumption is a concern.

- **FPGAs' role in LLM deployment**:

    - **Custom hardware logic**: Unlike CPUs and GPUs, FPGAs do not have a fixed hardware structure. This means that the logic gates within the device can be arranged to create custom hardware that is perfectly suited for specific LLM tasks, potentially offering superior performance for those tasks.

    - **Inference acceleration**: FPGAs can be particularly useful for accelerating inference in LLMs. Their reconfigurability allows them to be optimized for the precise operations of a deployed model, which can result in faster response times for applications requiring real-time processing.

    - **Edge computing**: FPGAs are also well-suited for deployment in edge devices. Their reconfigurability and efficiency make them ideal for situations where models need to be adjusted based on data being processed locally, and where power and space are limited.

    - **Integration with other technologies**: FPGAs can be used in conjunction with other accelerators, such as GPUs and TPUs, with each handling the tasks for which they are most suited. This can lead to a highly efficient heterogeneous computing environment.

Emerging technologies

Emerging technologies are pushing the boundaries of computational capability and efficiency, which can have profound implications for the development and deployment of LLMs. Let's take a closer look at some of these technologies.

ASICs (Application-Specific Integrated Circuits)

In the context of LLMs, ASICs are integrated circuits customized for a specific use, rather than for general-purpose use. The following are relevant regarding LLMs and ASICs:

- **Performance**: ASICs can provide performance optimizations specifically tailored to the computational patterns of LLMs, such as the matrix multiplications and nonlinear operations that are frequently used in these models

- **Energy efficiency**: ASICs are often more energy-efficient for the tasks they are designed for, which can be a significant advantage when deploying LLMs at scale, as energy costs can be a substantial part of the total cost of ownership

- **Cost**: While the initial design and manufacturing costs can be high, the per-unit cost of ASICs may be lower in the long term, especially when produced at scale

Neuromorphic computing

In neuromorphic computing, electronic analog circuits equipped systems are used to emulate the neuro-biological structures inherent in the nervous system. For LLMs, this could mean the following:

- **Parallel processing**: Similar to the brain, neuromorphic chips can handle many processes in parallel, potentially offering a different approach to handling the parallelism inherent in LLMs

- **Power consumption**: Neuromorphic chips can dramatically reduce power consumption, an important consideration when deploying LLMs in environments where power is limited, such as mobile devices or embedded systems

- **Real-time processing**: Neuromorphic chips might be particularly well-suited to applications that require real-time processing capabilities, such as natural language interaction in robotics

Quantum computing

To perform computation, quantum computing utilizes quantum-mechanical phenomena, such as superposition and entanglement, and holds promise for LLMs in several ways:

- **Speed**: Quantum computers may solve certain types of problems much faster than the best current classical computers, especially those involving complex optimizations and calculations, which are often part of LLM training and operations

- **New algorithms**: They could enable the development of new algorithms for LLMs that are not feasible on classical computers, potentially leading to breakthroughs in machine learning

- **Data handling**: The ability to handle massive datasets and perform computations on them in ways that classical computers cannot could revolutionize the way that LLMs are trained and used

Optical computing

Optical computing uses photons produced by lasers or diodes for computation. For LLMs, this could offer several benefits:

- **Speed**: Since light can travel faster than electrical signals, optical computing has the potential to perform computations at a much higher speed

- **Parallelism**: Light beams can travel through each other without interference, which could potentially allow for a high degree of parallelism in computations

- **Heat**: Optical computing generates less heat than electrical computing, addressing one of the major challenges in scaling up computational resources for LLMs

Each of these emerging technologies carries the potential to change the landscape of LLM deployment significantly. While some, such as ASICs, are already being used to some extent, others remain largely experimental and will require more development before they can be integrated into mainstream LLM applications. Nonetheless, they represent exciting prospects for the future of AI and computing in general.

System-level optimizations

System-level optimizations are critical for maximizing the performance and efficiency of LLMs. These optimizations span across the architecture and deployment strategies of computing resources. Here's a detailed look at the mentioned optimization strategies:

- **Distributed computing**:

    - **Parallel processing**: By spreading the computational workload of LLMs across multiple machines or nodes in a distributed system, each node can process a subset of data or a different part of a model simultaneously. This parallel processing can dramatically reduce the time required for tasks such as model training and inference.

    - **Resource scaling**: Distributed computing allows for the scaling of resources to match the demands of a workload. During periods of high demand, additional nodes can be added to a distributed system to maintain performance without requiring permanent investment in additional infrastructure.

    - **Fault tolerance**: Systems can be designed to handle node failures gracefully. If one node goes down, others can take over its workload without interrupting the overall operation of an LLM.

- **Heterogeneous computing**:

    - **Task-specific accelerators**: Different types of tasks required by LLMs may be best suited to different types of hardware accelerators. For example, GPUs can be used for parallel matrix operations, TPUs can be used for tensor operations, and FPGAs can be used for custom-designed logic that is optimized for specific tasks.

    - **Resource optimization**: A heterogeneous environment allows for each task to be routed to the most efficient processor for that task, optimizing both performance and energy consumption.

    - **Flexibility and adaptability**: Heterogeneous computing environments can be adapted to the changing needs of LLMs. As models and algorithms evolve, the computing environment can be reconfigured to best support the new requirements.

- **Edge computing**:

    - **Latency reduction**: By processing data closer to where it is generated or used, edge computing can significantly reduce latency, which is beneficial for applications that require real-time interaction, such as virtual assistants and real-time language translation.

    - **Bandwidth optimization**: Processing data on the edge can reduce the amount of data that needs to be transmitted over a network, conserving bandwidth and potentially reducing costs.

    - **Power and thermal management**: Edge devices often have strict constraints on power consumption and heat generation. Edge-specific accelerators are designed to operate within these constraints, ensuring that the devices can run LLMs without overheating or draining their power sources too quickly.

- **Data privacy and security**: Processing sensitive data on the edge can enhance privacy and security by minimizing the transmission of data to central servers, which can be particularly important for compliance with data protection regulations.

Advanced hardware acceleration techniques for LLMs are not solely about raw computational power; they are also about efficiency, adaptability, and the ability to integrate seamlessly with software frameworks. As the field of machine learning continues to evolve, so too will the hardware that supports it, leading to continuous improvements in the speed, cost, and capability of LLMs.

Efficient data representation and storage

Efficient data representation and storage in the context of LLMs extends beyond quantization and pruning to encompass a variety of techniques and strategies. These approaches aim to reduce a model's memory footprint and speed up computation, which are crucial for storage limitations and quick data retrieval. Let's take a detailed look at advanced methods for efficient data representation and storage:

- **Model compression**:

  - **Weight sharing**: Reduces the model size by having multiple connections in the neural network share the same weight, effectively reducing the number of unique weights that need to be stored

  - **Sparse representations**: Beyond pruning, employing formats specifically designed for storing sparse matrices (such as CSR or CSC) can dramatically reduce the memory needed to store weights that are predominantly zeros

  - **Low-rank factorization**: Decomposes weight matrices into smaller, lower-rank matrices that require less storage space and can be recombined for computations

  - **Parameter sharing**: Across different parts of a model or between multiple models, parameters can be shared to reduce redundancy, especially in models with repetitive or recursive structures

  - **Tensor decomposition**: A technique that breaks down multidimensional arrays (tensors) into lower-dimensional components to reduce storage requirements, while maintaining computational efficiency

- **Optimized data formats**:

  - **Fixed-point representation**: Instead of using floating-point representations, which require more storage space and bandwidth, fixed-point numbers can be used to store weights and activations, significantly reducing the model size

  - **Binarization**: In extreme cases, weights and activations within neural networks can be binarized (reduced to ones and zeros), which can massively reduce the storage requirements and speed up computation by using bitwise operations

- **Memory optimization techniques**:

  - **Checkpointing**: During training, instead of storing all intermediate activations for backpropagation, only a subset is stored, and the rest are recomputed during the backward pass, trading computational time for memory

  - **In-place operations**: Modifying data directly in memory without creating copies can save memory bandwidth and storage

- **Efficient algorithms for storage and retrieval**:

  - **Data deduplication**: Involves eliminating duplicate copies of repeating data, which can be particularly effective in datasets with significant redundancy

  - **Lossless data compression**: Algorithms such as Huffman coding or arithmetic coding can compress data without losing information, making the storage and retrieval processes more efficient

- **Software-level optimizations**:

  - **Memory-efficient data structures**: Using advanced data structures that use memory more efficiently, such as tries for word storage in NLP tasks

  - **Optimized serialization**: When storing or transmitting model parameters, using efficient serialization formats can reduce the size of the data payload

- **Custom storage solutions**:

  - **Custom file systems**: Tailoring or using specialized filesystems that are optimized for the specific access patterns of LLMs, which can result in faster data retrieval times and better utilization of available storage

  - **Distributed storage systems**: Utilizing distributed filesystems that can scale horizontally and manage data across multiple nodes efficiently, thus enhancing data access and processing speed

Incorporating these advanced techniques requires careful planning and a deep understanding of both the models and the hardware on which they are run. The goal is to maintain, or even enhance, a model's ability to learn and make predictions while reducing the computational load and storage space required. The choice of which techniques to apply will depend on the specific constraints and requirements of the deployment environment, as well as the nature of the LLM being used.

Speeding up inference without compromising quality

Speeding up inference while maintaining quality is a key challenge in deploying LLMs effectively, especially in real-time applications. The techniques mentioned, distillation and optimized algorithms, are just part of a broader suite of strategies that can be employed to this end. Let's take a deeper dive into these and other methods.

Distillation

Distillation in the context of machine learning, particularly for LLMs, is a technique that helps in transferring knowledge from a larger, more complex model to a smaller, more efficient one. This process not only makes a model more deployable but also often retains a significant amount of the larger model's accuracy. Let's take an in-depth look at the various distillation techniques:

- **Soft target distillation**:

  - **Knowledge transfer**: Soft target distillation transfers the "knowledge" encoded in the probability distributions of a larger model's outputs to a smaller model. Instead of just learning from the ground truth labels (that is, hard targets), the smaller model learns to mimic the output distributions (that is, soft targets) of the larger model.

  - **Rich information**: The soft targets provide a richer set of information compared to hard targets, which can include insights into the confidence of a model's predictions and the relationships between different classes.

  - **Improved generalization**: By training on these soft targets, the smaller model can capture the nuanced decision-making process of the larger model, leading to better generalization from the same training data.

- **Intermediate layer distillation**:

  - **Layer activations**: This method involves using the activations from the intermediate layers of the larger model as additional training signals for the smaller model. These activations represent higher-level features that the larger model has learned to extract from data.

  - **Enhanced feature learning**: By aiming to replicate these intermediate representations, the smaller model can potentially learn a similar feature hierarchy, which can be especially valuable for complex tasks that require a deep understanding of the input data.

  - **Preserving model capabilities**: Intermediate layer distillation is particularly useful to ensure that the distilled model preserves the capabilities of the larger model, including the ability to represent and process data in sophisticated ways.

- **Attention distillation**:

  - **Attention mechanisms**: Attention mechanisms in models, particularly those based on the Transformer architecture, allow a model to weigh the importance of different parts of the input data when making predictions.

  - **Transferring focus**: Attention distillation focuses on transferring these attention patterns from the larger model to the smaller one. This means that the smaller model learns not just what to predict but also where to focus its computational resources.

- **Preserving contextual understanding**: Attention patterns are crucial for tasks that require an understanding of context and relationships within data. Distilling these patterns helps the smaller model maintain a similar level of contextual awareness as the larger model.

Distillation techniques are particularly useful in deploying LLMs in resource-constrained environments, such as mobile devices, edge computing nodes, or any situation where the computational resources are limited. They offer the benefits of introducing large, highly accurate models in scenarios where it would otherwise be impractical to deploy them directly. Through these techniques, models can be made more efficient without a substantial loss in performance, making AI more accessible and versatile.

Optimized algorithms

Optimized algorithms are essential for enhancing the efficiency of LLMs, particularly during the inference phase when a model is used to make predictions or generate text. Let's delve into the specifics of efficient inference algorithms and algorithmic simplifications:

- **Efficient inference algorithms**:

  - **Approximate Nearest Neighbor (ANN) search**: In tasks such as retrieval-based question answering or document retrieval, where the goal is to find the most similar items from a large dataset, exact nearest neighbor searches can be prohibitively slow. ANN algorithms, such as **Locality-Sensitive Hashing** (LSH), tree-based methods such as KD-trees, or graph-based approaches such as **Hierarchical Navigable Small World** (HNSW) graphs, provide a way to quickly find a "good enough" match without exhaustively comparing every possible item.

  - **Sublinear time complexity**: Many efficient inference algorithms are designed to have sublinear time complexity with respect to the size of the data they process, meaning that the time they take to execute does not increase linearly with the size of the dataset.

- **Algorithmic simplifications**:

  - **Beam search**: For generative tasks such as translation or summarization, beam search is a common technique used instead of an exhaustive search. It limits the number of possibilities considered at each step of the generation process to the "best" few, according to a scoring function. This reduces the number of computations needed to generate an output sequence while still maintaining high-quality results.

  - **Greedy decoding**: In some cases, even simpler than beam search, greedy decoding takes only the most probable next step at each point in a sequence without considering multiple alternatives. This can be significantly faster and is often used in scenarios where speed is more critical than achieving the absolute best performance.

  - **Quantization and pruning**: These techniques can also be considered a form of algorithmic optimization. By reducing the precision of the computations (quantization) or the number of parameters in the model (pruning), inference can be performed more quickly.

- **Customized algorithms for specific tasks**:

  - **Tailored algorithms**: Algorithms can be tailored to the specific characteristics of the tasks that an LLM is designed for. For instance, if the LLM is mostly used for tasks that don't require understanding the full complexity of language, such as simple classification, then the inference algorithms can be simplified accordingly.

  - **Algorithm adaptation**: Existing algorithms can be adapted to make use of the hardware acceleration features available, such as the tensor cores in GPUs. This involves rewriting the algorithms to leverage parallelism and specialized computational units effectively.

- **Benefits of optimized algorithms**:

  - **Increased throughput**: By reducing the time it takes to perform inference, more requests can be processed in the same amount of time, increasing the overall throughput of the system

  - **Lower resource usage**: Faster inference generally means less computational resource usage, which can reduce operating costs, especially in cloud-based environments

  - **Enabling real-time applications**: Efficient algorithms are critical for applications that require real-time responses, such as conversational AI, where delays in response times can degrade the user experience

In summary, optimized algorithms play a critical role in the practical deployment of LLMs. They help balance the computational demands of these models with the need for speed and efficiency, enabling their use in a wider range of applications and making them more accessible for users and businesses alike.

Additional methods

In the domain of machine learning, and especially in the application of LLMs, various additional methods can be employed to enhance performance and efficiency at inference time. These methods are designed to optimize the computational demands of LLMs, allowing them to operate more swiftly and effectively on a wide range of hardware. A detailed exploration of these techniques is as follows:

- **Model quantization**:

  - **Reduced precision**: As discussed in the previous chapter, quantization involves lowering the precision of a model's computations from floating-point representations (such as 32-bit floats) to lower-bit representations (such as 8-bit integers), which can significantly speed up inference times

  - **Hardware compatibility**: Many modern processors, especially those designed for mobile devices, are optimized for low-precision arithmetic, making quantization an effective method to improve performance on such devices

- **Layer fusion:**

  - **Optimized computation**: Layer fusion combines the operations of multiple layers into a single operation. This can reduce the computational overhead and memory access required for separate layers, thus decreasing inference latency.

  - **Streamlined processing**: By fusing layers, the amount of data that needs to be moved between different stages of a model is reduced, leading to faster processing times.

- **Cache mechanisms:**

  - **Result reuse**: Caching involves storing the results of computations so that if the same computation is needed again, the result can be retrieved from the cache rather than being recalculated

  - **Intermediate computation storage**: Caching can also apply to intermediate computations within an LLM, which is beneficial when similar inputs are processed repeatedly

- **Early exiting:**

  - **Confidence-based termination**: Some models can be structured to allow for an early exit if a model is sufficiently confident in its prediction. This means the inference process can be truncated, saving computational resources.

  - **Layer-wise confidence checking**: Early exiting typically involves checking the confidence of the prediction at various points in a model and exiting if certain criteria are met.

- **Hardware-specific optimizations:**

  - **Tailored models**: Optimizing models for specific types of hardware can involve tweaking the architecture of the model or the implementation of the algorithms to take full advantage of the hardware's capabilities

  - **Instruction set utilization**: Different processors have different instruction sets and capabilities, and optimizing models to leverage these can lead to better performance

- **Parallelization of inference tasks:**

  - **Concurrent processing**: Parallelization involves spreading out the inference workload across multiple processing units, which can be particularly effective on GPUs and multi-core CPUs

  - **Task distribution**: Tasks can be distributed across processors in a way that minimizes data transfer and maximizes the use of available computational resources

- **Network pruning and sparsity**:

  - **Redundant weight removal**: As discussed in the previous chapter, pruning involves removing weights from a network that contribute little to the output, leading to a sparser and more efficient network

  - **Sparsity-induced speed**: Sparse models often require fewer operations to achieve the same result, leading to faster inference times, especially on hardware that can exploit sparsity for performance gains

In summary, speeding up inference without compromising quality encompasses a variety of techniques, from model-specific strategies such as distillation to algorithmic and system-level optimizations. These strategies are often complementary, and a combination of them can be used to meet the specific performance needs of an application. The choice of technique will depend on the particular LLM, the hardware platform, the nature of the task, and the required balance between speed and accuracy.

Balancing cost and performance in LLM deployment

Balancing the cost and performance in LLM deployment is a multifaceted challenge that involves a strategic approach to infrastructure and resource management. Let's explore a detailed exploration of the elements.

Cloud versus on-premises

Choosing between cloud and on-premises solutions to deploy LLMs involves weighing the pros and cons of each in terms of scalability, cost, operational overhead, data security, and customization. Here is a more detailed exploration of these considerations:

- **Scalability**:

  - **Cloud**: Cloud platforms offer dynamic scalability, allowing organizations to increase or decrease their computational resources in response to their needs. For LLM workloads that are not constant, this means not having to pay for unused resources during off-peak times, as well as the ability to handle surges in demand without the risk of service degradation.

  - **On-premises**: Scaling on-premises infrastructure typically requires purchasing additional hardware, which may lead to underutilized resources during periods of low demand. However, for organizations with predictable and constant high demand, on-premises solutions can be more stable and predictable in performance.

- **Initial investment**:

  - **Cloud**: Typically operates on a pay-as-you-go model, reducing the need for large initial investments. Organizations can start deploying LLMs without committing to large expenditures on hardware and data center space.

  - **On-premises**: Requires significant capital expenditure for the purchase of servers, storage, networking equipment, and the infrastructure needed to house and maintain them. This investment makes more sense for organizations that need resources consistently over time.

- **Operational overheads**:

  - **Cloud**: The cloud service provider manages the maintenance of the infrastructure, including updates and repairs, which can reduce the need for specialized IT staff within an organization and potentially lower operational costs.

  - **On-premises**: Organizations are responsible for the ongoing maintenance and updating of their infrastructure, which can be costly and require a dedicated IT team.

- **Data sovereignty and privacy**:

  - **Cloud**: While cloud providers generally offer robust security features, there may still be concerns around data sovereignty and privacy, especially when sensitive data is stored or processed in the cloud.

  - **On-premises**: Offers more control over data security because data remains within an organization's controlled environment. This can be crucial for compliance with data protection regulations and for organizations that handle particularly sensitive information.

- **Customization**:

  - **Cloud**: While cloud services offer a range of options and configurations, there may be limitations in terms of the hardware and software stacks available, which could impact the performance of LLMs that have specific requirements

  - **On-premises**: Allows organizations to tailor their infrastructure precisely to their needs, optimizing both the hardware and software environment for their specific LLM workloads, which can lead to better performance

- **Deciding factors for LLM deployment**:

  - **Cost-benefit analysis**: Organizations must conduct a thorough cost-benefit analysis to determine which model offers the best value for their specific use case

  - **Technical requirements**: The technical demands of the LLMs in question, such as processing power, memory, and storage, will significantly influence the decision

- **Long-term strategy**: The choice between cloud and on-premises should align with an organization's long-term strategy, considering factors such as anticipated growth, technological developments, and budgeting

Model serving choices

When it comes to deploying LLMs, the infrastructure used to serve the models to end users or applications is a critical factor. There are several model serving choices, each with its own set of advantages and potential drawbacks. Let's explore these options in detail:

- **Dedicated servers**:

  - **Robust performance**: Dedicated servers provide powerful and consistent performance because they are not shared with other services or applications. They can be fully utilized by an LLM, ensuring that the maximum computational resources are available when needed.

  - **Customization**: They allow for deep customization and tuning of the hardware and software environment, which can lead to significant performance improvements for specific LLM workloads.

  - **Potential for underutilization**: One downside is the potential for resource underutilization during periods of low demand. This can make dedicated servers less cost-effective, especially if the demand for an LLM is variable.

- **Serverless architectures**:

  - **Cost-efficiency**: Serverless architectures abstract away the server management and automatically scale to match demand. This means you pay only for the compute time you consume, without having to maintain idle servers during downtime.

  - **Flexibility**: They offer great flexibility and are ideal for unpredictable or fluctuating workloads, as the infrastructure can quickly adapt to changes in usage patterns.

  - **Performance constraints**: However, serverless architectures may impose limitations on the maximum runtime of functions and the resources available to them, which could affect performance, especially for compute-intensive LLM tasks.

- **Containerization**:

  - **Portability**: Containerization, using technologies such as Docker and Kubernetes, allows an LLM to be packaged with all its dependencies, ensuring consistent behavior across different computing environments.

  - **Scalability and control**: Containers strike a balance between the scalability offered by cloud services and the control provided by on-premises servers. They can be easily scaled up or down, based on demand.

- **Resource efficiency**: Containers can be more resource-efficient than virtual machines, as they share the host system's kernel and avoid the overhead of simulating an entire operating system.

- **Other considerations**:

  - **Latency**: For interactive applications that use LLMs, such as virtual assistants or chatbots, the latency in response times can be a crucial factor. Dedicated servers often provide the lowest latency, but modern container orchestration and serverless platforms have made significant strides in reducing latency as well.

  - **Maintenance and upkeep**: With dedicated servers and containerized environments, there's a need for ongoing maintenance and updates, which can be handled by the cloud service provider in serverless architectures.

  - **Security and compliance**: Depending on the nature of the data that is processed by an LLM and the regulatory environment, security and compliance requirements may influence the choice of infrastructure.

Cost-effective and sustainable deployment

Cost-effective and sustainable deployment of LLMs is critical for organizations looking to harness the power of advanced AI without incurring prohibitive costs. Let's take a comprehensive look at the strategies to achieve this balance:

- **Hardware acceleration**:

  - **Performance versus cost**: Specialized hardware such as GPUs, TPUs, and FPGAs can significantly accelerate LLM operations. GPUs are widely used for their parallel processing capabilities, TPUs are optimized for tensor operations, and FPGAs offer customizable logic for specific tasks. However, these come with varying price tags and operational costs, and the decision to use one over the others will depend on the specific computational needs of the LLM tasks, as well as budget limitations.

  - **Efficiency**: The efficiency of hardware accelerators can also affect costs. More efficient hardware can process more data at a lower energy cost, which is an important consideration for long-term sustainability.

- **Data management**:

  - **Storage optimization**: Efficient data storage solutions are essential to handle the vast amounts of data processed by LLMs. Employing data compression and deduplication strategies can decrease the storage footprint.

  - **Caching mechanisms**: Implementing caching can significantly reduce I/O operations by storing frequently accessed data in a quickly accessible cache, thus reducing latency and lowering costs associated with data transfer and processing.

- **Computational strategies**:

  - **Model quantization**: As previously discussed, this involves reducing the precision of model parameters and computations, which can lead to faster computation and reduced model size, making LLMs less expensive to run and easier to deploy on edge devices

  - **Pruning**: By removing non-critical parts of a neural network, pruning can simplify a model, reducing its computational requirements and, therefore, the cost of running the model

  - **Distillation**: Training smaller models to mimic the performance of larger, more complex ones can make deployment more feasible, by using fewer computational resources without a significant drop in accuracy

- **Monitoring and optimization**:

  - **Performance tracking**: Continuous monitoring of both performance and costs can identify inefficiencies. Tools and platforms that offer real-time monitoring and alerting can be crucial in managing operational costs.

  - **Optimization**: Regular analysis of LLMs' performance data can reveal opportunities for optimization, such as fine-tuning configurations, updating models, or improving algorithms.

- **Elasticity and auto-scaling**: Cloud services often allow you to automatically scale resources up or down based on real-time demand. This elasticity means that organizations only pay for the compute and storage resources they actually use.

- **Life cycle management**:

  - **Holistic view**: Understanding the entire life cycle of LLMs, from initial development and training through to deployment and ongoing maintenance, can uncover areas where costs can be minimized. For example, training costs can be high, so optimizing the training process can lead to substantial savings.

  - **Continuous improvement**: As LLMs are used, they can generate new data that can be used to refine and improve them. Incorporating this new data can improve efficiency and reduce the need for costly retraining from scratch.

In conclusion, organizations aiming to deploy LLMs must navigate these factors to strike a balance between computational power and cost efficiency. This includes making informed decisions about infrastructure, considering both immediate needs and future scalability, and selecting serving architectures that align with usage patterns and performance requirements. Ultimately, the right mix of technology and strategy can lead to a sustainable and cost-effective LLM deployment.

Summary

Advanced hardware acceleration techniques provide pivotal enhancements to the capabilities of LLMs, by significantly boosting the speed and efficiency of computations required for their training and inference phases. This acceleration is largely achieved through the integration of specialized hardware components and architectural innovations in modern GPUs, as well as the strategic application of various computational methodologies.

Tensor cores, a feature of contemporary GPUs, greatly expedite matrix operations crucial to deep learning by enabling mixed-precision arithmetic—utilizing both FP16 and FP32 formats to balance computational speed with precision. This capability not only accelerates matrix multiplications but also increases the overall throughput for deep learning tasks, leading to more rapid model training and quicker inference.

Optimization of memory hierarchy is another critical area. Advanced GPUs optimize the usage of shared, cache, and global memory types, which is fundamental for reducing data movement – a common performance bottleneck. High bandwidth memory technologies such as GDDR6 and HBM2 further enhance the data transfer rates, enabling more efficient processing of the large datasets that are typical in LLM applications.

The asynchronous execution capabilities of GPUs, such as concurrent kernel execution and overlapping of data transfer with computation, ensure maximum utilization of computational units, thereby minimizing latency and improving performance. By facilitating multiple operations simultaneously through their multiple stream processors, GPUs can efficiently manage various execution tasks in parallel, significantly boosting the efficiency of LLM operations.

These advancements collectively result in GPUs that are not only faster but also smarter in managing computations and data flow. This is particularly important in the field of deep learning, where processing vast volumes of data expeditiously is often crucial to the feasibility of deploying sophisticated AI solutions. By leveraging these advanced features, developers and researchers can train more complex models, accelerate experimentation, and deploy more advanced AI systems, ultimately pushing the frontiers of what's achievable with generative AI.

In the next chapter, we move on to review LLM vulnerabilities, bias, and legal implications.

Part 4:
Issues, Practical Insights, and Preparing for the Future

In this part, you will learn about identifying and mitigating risks, confronting biases in LLMs, legal challenges in LLM deployment and usage, regulatory landscape and compliance, and ethical considerations. We will provide you with business case studies from which you will learn the concept of ROI. Additionally, you will see a survey of the landscape of AI tools, a comparison between open source and proprietary tools, an explanation of how to integrate LLMs with existing software stacks, and an exploration of the role of cloud providers in NLP. You will learn about what to expect from the next generation of LLMs and how to get ready for GPT-5 and beyond. We will conclude with key takeaways from this guide, the future trajectory of LLMs in NLP, and final thoughts about the LLM revolution.

This part contains the following chapters:

- *Chapter 11, LLM Vulnerabilities, Biases, and Legal Implications*
- *Chapter 12, Case Studies – Business Applications and ROI*
- *Chapter 13, The Ecosystem of LLM Tools and Frameworks*
- *Chapter 14, Preparing for GPT-5 and Beyond*
- *Chapter 15, Conclusion and Looking Forward*

11

LLM Vulnerabilities, Biases, and Legal Implications

In this chapter, we will explore the complexities surrounding LLMs, focusing on their vulnerabilities and biases. We will discuss the impact of these issues on LLM functionality and the efforts needed to mitigate them. Additionally, we will provide an overview of the legal and regulatory frameworks governing LLMs, highlighting intellectual property concerns and the evolving global regulations. We will aim to balance the perspectives on technological advancement and ethical responsibilities in the field of LLMs, emphasizing the importance of innovation aligned with regulatory caution. We will end the chapter with a case study regarding bias mitigation.

In this chapter, we're going to cover the following main topics:

- LLM vulnerabilities – identifying and mitigating risks

- Confronting biases in LLMs

- Legal challenges in LLM deployment and usage

- Regulatory landscape and compliance for LLMs

- Ethical considerations and future outlook

- Hypothetical case study – bias mitigation in AI for hiring platforms

By the end of this chapter, you should possess a comprehensive understanding of the multifaceted challenges associated with LLMs, ranging from vulnerabilities and biases to legal and regulatory complexities.

LLM vulnerabilities – identifying and mitigating risks

The deployment and usage of LLMs bring forward significant challenges and considerations in the domains of security, ethics, law, and regulation. LLM vulnerabilities need to be thoroughly identified and mitigated to protect these systems from potential abuses or malfunctions, which can stem from adversarial attacks or unintended model behaviors. Developers must implement robust security protocols and continually monitor for vulnerabilities that could compromise the integrity or performance of LLMs.

LLMs are susceptible to a range of vulnerabilities that can impact their integrity, performance, and reliability. Here are some detailed considerations.

Identification of security risks

The identification of security risks in LLMs is a critical step in safeguarding their integrity and ensuring they function as intended. Let's take a closer look at the process and why it's important:

- **Adversarial attacks**:

  - LLMs can be susceptible to adversarial attacks, where input data is intentionally manipulated to cause the model to make mistakes or produce incorrect outputs. These attacks exploit weaknesses in the model's understanding of the input data.

  - To counter such threats, LLMs must be rigorously tested against potential adversarial inputs. This involves not only traditional validation methods but also crafting and testing against inputs designed to deceive the model.

- **Vulnerability scanning and testing**:

  - Regular scans and tests of LLMs are necessary to identify new vulnerabilities that could emerge as the models are exposed to new data or as attackers develop new strategies.

  - Automated tools can scan for known types of vulnerabilities, but it's also essential for security experts to conduct creative testing to discover unknown weaknesses.

- **Proactive security measures**:

  - Beyond identifying risks, it's important to implement measures that can proactively prevent attacks or minimize their impact. This might include input validation, anomaly detection mechanisms, and regular updates to the model as new threats are identified.

- **Continuous security monitoring**:

  - Security is not a one-time task but a continuous process. As LLMs learn and evolve, their threat landscape may change, necessitating ongoing monitoring and re-assessment of risks.

- **Collaborative efforts**:

  - Sharing information about threats and defenses within the community can help in developing robust security practices. Collaboration between researchers, developers, and security professionals can lead to the creation of more secure systems.

Mitigation strategies

Mitigation strategies for security risks in LLMs involve a proactive and multifaceted approach to prevent, detect, and respond to potential threats. Here's an in-depth explanation of the strategies mentioned:

- **Robust security protocols**:

  - **Input validation**: To prevent adversarial attacks, it's crucial to validate the inputs to LLMs. This means ensuring that the data fed into the model conforms to expected patterns and is free from malicious manipulations designed to deceive the model.

  - **Anomaly detection**: Anomaly detection systems can identify unusual patterns in data processing that may signify an attempt to exploit model vulnerabilities. These systems use statistical models to establish a baseline of normal activity and flag deviations from this baseline for further investigation.

  - **Data encryption**: Encrypting data both in transit to and from the model, as well as at rest, secures the inputs and outputs against interception and tampering. This helps in maintaining the confidentiality and integrity of the data being processed by the LLM.

- **Comprehensive monitoring system**:

  - **Performance tracking**: A system that continuously monitors the LLM's performance can detect sudden changes that might indicate an issue, such as a drop in accuracy that could result from an attack.

  - **Behavior analysis**: Monitoring the behavior of LLMs can help in understanding how they respond to different inputs. Abnormal behavior patterns can be early indicators of security issues.

  - **Alerting mechanisms**: The system should be capable of generating alerts when potential vulnerabilities are detected, enabling developers and security teams to take immediate action to investigate and remediate the issue.

  - **Failure detection**: In addition to security threats, monitoring systems can also detect failures in the model that could affect its reliability, prompting preventative maintenance or updates to the model to ensure it continues to operate correctly.

Continual learning and updates

Continual learning and updates in the context of LLMs are multifaceted and revolve around several core principles aimed at maintaining efficacy and security over time.

Continual learning in LLMs

Continual learning is the capacity of an AI system to gradually assimilate new data while retaining previously learned information. This is crucial because the world is dynamic; new information emerges, and language evolves. For instance, new slang terms, neologisms, or even entirely new dialects may develop. An LLM that can't incorporate new language use would quickly become outdated.

In practice, continual learning might involve techniques such as the following:

- **Online learning**: Where the model updates its parameters on the fly as new data comes in

- **Transfer learning**: Adapting a pre-trained model to new tasks or datasets with additional training

- **Meta-learning**: Sometimes called "learning to learn," where the model is trained on a variety of tasks in such a way that it can quickly adapt to new, unseen tasks with minimal additional data

Continual learning poses technical challenges, such as avoiding catastrophic forgetting (where learning new information causes the model to forget old information) and ensuring that updates do not introduce biases or reduce the model's performance on previous tasks. Techniques on how to deal with these technical challenges are included in several other chapters of this book.

Updates for performance

Aside from learning new data, LLMs need to be updated to improve performance. This could involve architectural changes that allow the model to process information more efficiently or updates to the training process to produce more accurate outputs. For instance, if users frequently ask about AR and VR technologies, the model might be updated to have a deeper understanding of these topics, providing more detailed and accurate responses.

Security updates

Security is another significant aspect of updates. As cyber threats evolve, models must be hardened against them. Here's why it's crucial:

- **Data integrity**: Ensuring that the data used for training is free of tampering or corruption

- **Model robustness**: Protecting against adversarial attacks, where inputs are designed to trick the model into making errors

- **Privacy**: Updating mechanisms to protect sensitive information, especially as models are increasingly able to understand and generate natural language content that could contain personal data

Regular patching with security enhancements means not just updating the software that interfaces with the LLM but sometimes altering the model itself. For instance, if a vulnerability is found that allows an attacker to extract data from the model, the model may need to be retrained to resist this type of attack.

The process of updating LLMs

Updates to LLMs involve a cycle of monitoring, development, testing, and deployment:

1. **Monitoring**: Continuously checking the model's performance and watching for emerging threats and opportunities for improvement.

2. **Development**: Creating updates, whether they're new training routines, architectural changes, or security patches.

3. **Testing**: Rigorously evaluating updates in controlled environments to ensure they don't degrade the model's performance or security.

4. **Deployment**: Rolling out the update, which could be done incrementally or all at once, depending on the nature of the update and the operational requirements of the LLM.

Collaboration with security experts

Collaboration with security experts is a strategic approach to safeguarding LLMs against a multitude of potential threats. Cybersecurity experts are at the forefront of understanding the latest threats. By collaborating with these experts, developers of LLMs can gain the following:

- **Threat intelligence**: Security experts often have access to the latest intelligence about potential cyber threats, including those from state actors, cybercriminals, and other malicious entities

- **Predictive analysis**: Through the use of advanced threat modeling and predictive analytics, experts can forecast potential vulnerabilities and attack vectors that might be exploited in the future

Development of best defense strategies

Cybersecurity experts help in developing robust defense mechanisms using the following:

- **Tailored defense mechanisms**: Designing specific security measures that address the unique needs of LLMs, such as securing the data pipelines, preventing unauthorized access, and protecting against data poisoning attacks

- **Incident response planning**: Creating detailed plans for how to respond to security breaches, which is critical for minimizing damage and restoring normal operations as quickly as possible

- **Involvement in design and deployment**: Incorporating security experts during the design and deployment phases of LLMs can lead to the following:

  - **Secure-by-design principles**: Embedding security into the architecture of LLMs from the very beginning, which can reduce the risk of vulnerabilities and make the systems more resilient to attacks

- **Security audits**: Conducting thorough security audits throughout the design and deployment processes to identify and rectify any weaknesses

- **Built-in protections**: With expert involvement, LLMs can be equipped with a variety of built-in protections:

  - **Data encryption**: Implementing strong encryption standards for both at-rest and in-transit data to prevent unauthorized access or leaks

  - **Authentication protocols**: Using robust authentication mechanisms to ensure that only authorized individuals can access the LLMs

  - **Regular security patches**: Establishing a routine for applying security patches to protect against known vulnerabilities

  - **Redundancy and fail-safes**: Designing systems with redundancy to prevent single points of failure and implementing fail-safe mechanisms to maintain essential functions even under duress

- **Continuous collaboration**: Effective cybersecurity measures for LLMs include the following:

  - **Training and awareness**: Ensuring that all stakeholders, from developers to end users, are trained in basic security awareness and best practices

  - **Community engagement**: Participating in cybersecurity communities to stay abreast of new developments, share knowledge, and collaborate on solutions to emerging threats

  - **Compliance and standards**: Working with experts to ensure that LLMs comply with relevant laws, regulations, and industry standards related to cybersecurity

Ethical hacking and penetration testing

Ethical hacking and penetration testing are proactive security measures critical to the defense strategy of any technological system, including LLMs. They are particularly important in the rapidly evolving digital world where new vulnerabilities can be exploited by malicious actors.

- **Ethical hacking**: Ethical hacking involves employing cybersecurity experts who are authorized to identify and exploit vulnerabilities in systems. The key aspects include the following:

  - **Authorized testing**: Ethical hackers have permission to probe the system's defenses, which differentiates their activities from malicious hacking.

  - **Skill utilization**: Ethical hackers typically possess the same technical skills as malicious hackers but use these skills to improve security rather than to exploit vulnerabilities.

  - **Vulnerability identification**: They actively search for weaknesses in a system, such as susceptibility to SQL injection, cross-site scripting, or other types of attacks that could compromise LLMs.

- **Reporting and remediation**: After identifying vulnerabilities, ethical hackers report them to the organization. This allows the organization to address the issues before they can be exploited by attackers.

- **Penetration testing**: Penetration testing, or pen testing, takes a structured approach to finding security weaknesses with the help of the following:

  - **Simulated attacks**: Pen tests simulate real-world attacks on systems to identify vulnerabilities that could be exploited by attackers

  - **Comprehensive evaluation**: The testing covers numerous aspects of the system, including network infrastructure, applications, and end-user behaviors

  - **Testing methodologies**: There are different types of penetration tests, including black-box (with no prior knowledge of the system), white-box (with full knowledge), and gray-box (with partial knowledge), each providing different insights into system security

  - **System hardening**: The insights from penetration testing are used to harden systems against attacks by fixing the vulnerabilities found and improving the overall security posture

- **Regular and iterative process**: A regular and iterative process for LLMs includes the following:

  - **Regular scheduling**: Regularly scheduled tests are essential as new vulnerabilities can emerge at any time due to changes in the system, updates, or the discovery of new hacking techniques.

  - **Adapting to new threats**: As LLMs evolve, so do the threats against them. Continuous testing ensures that defenses are always based on the latest threat intelligence.

  - **Compliance and trust**: These practices not only help to secure systems but also play a role in regulatory compliance and building trust with users by demonstrating a commitment to security.

Ensuring the security of LLMs is a dynamic and ongoing process that requires vigilance, expertise, and a proactive approach to risk management. As LLMs become more widespread, the importance of securing them against adversarial attacks and malfunctions grows in tandem, demanding a consistent and dedicated effort from AI developers and security professionals.

Confronting biases in LLMs

Confronting biases in LLMs is a critical challenge within the field of AI. These biases can manifest in various forms, often reflecting and perpetuating the prejudices present in the training data. Addressing these biases is essential to build fair and equitable AI systems. Here's a more detailed exploration:

- **Careful dataset curation**:

  - The process begins with the selection and preparation of training datasets. Curators must ensure that the data is representative of diverse perspectives and does not contain discriminatory or biased examples. This might involve including data from a wide range of sources and demographic groups.

- Active efforts to identify and remove biased or offensive content from training datasets are crucial. This can be achieved through both automated filtering algorithms and human review.

- **Secure data handling**: Proper handling of data ensures it remains protected from unauthorized access throughout the curation process. Implementing strong security measures helps maintain the integrity and confidentiality of sensitive datasets used in training.

- **Access controls**: Limit access to sensitive training datasets through role-based access control, ensuring that only authorized personnel can view or modify the data.

- **Unbiased model training methodologies**:

  - Developing training methodologies that do not inherently favor one outcome over another is key. This includes designing algorithms that are sensitive to the potential for bias and that actively work to minimize it.

  - Techniques such as adversarial training, where the model is exposed to scenarios specifically designed to counteract biases, can be employed. Another method is regularization, which can discourage the model from relying too heavily on features associated with bias.

  - **Anonymization and de-identification**: Personal or sensitive data in the training set should be anonymized or de-identified to prevent exposing individual identities or demographic details that could introduce bias.

- **Consistent evaluation to ensure fairness in outcomes**:

  - Continuous evaluation of the model's outputs is necessary to monitor for biases. This involves testing the model against benchmarks designed to detect unfair or biased decision-making.

  - Implementing fairness metrics, which can quantitatively measure biases in model outputs, is an integral part of the evaluation process. These metrics can guide the ongoing development of the model to mitigate biases effectively.

- **Transparency and explainability**:

  - Building models that are transparent and explainable aids in identifying where and how biases may be occurring. If users and developers understand the reasoning behind a model's decisions, they can more easily spot biases.

  - Explainable AI frameworks can provide insights into the model's decision-making process, highlighting aspects of the data that are weighted more heavily and may contribute to biased outcomes.

- **Secure model deployment**: Once an LLM is ready for deployment, it's essential to ensure secure deployment practices. Secure model deployment ensures that the model runs in environments free from vulnerabilities, reducing the risk of biased manipulation or malicious usage.

- **Engagement with stakeholders**:

  - Collaboration with stakeholders, including those who may be affected by the model's decisions, can provide valuable insights into the potential impacts of biases. This can inform the development process and help prioritize efforts to address the most significant issues.

  - Diverse teams that include members from various backgrounds can also help anticipate and identify biases that might not be apparent to a more homogenous group.

In summary, confronting biases in LLMs is an ongoing process involving careful attention at every development stage, from dataset curation to evaluation. The goal is to create fair and equitable AI systems that benefit everyone and minimize harm, making it both a technical and ethical imperative.

Legal challenges in LLM deployment and usage

Addressing the legal challenges associated with the deployment and usage of LLMs is critical, as these systems increasingly affect various aspects of society and commerce. In this section, we will take a closer look at the two main legal areas.

Intellectual property rights and AI-generated content

The topic of **intellectual property (IP)** rights in the context of AI-generated content is complex and still an emerging area of law. The creation of content by LLMs raises several challenging questions regarding the ownership and control of IP. Here's an in-depth look into the different facets of this issue:

- **Ownership of AI-generated content**:

  - **Legal precedents**: Historically, IP law has been built around the idea of human authorship. AI challenges this notion because it can generate content independently after being initially programmed by humans.

  - **Human versus machine**: Most current legal frameworks do not recognize AI as an independent creator with the capacity to hold IP rights. Instead, they focus on human involvement in the creative process.

  - **Copyright**: The copyright status of AI-generated content is debated. Is the content an original work of authorship, which is a criterion for copyright protection, or is it merely the result of an algorithm processing data?

- **Stakeholders in IP rights**:

  - **Creators of the algorithms**: The developers of the AI may claim ownership, arguing that their software is the "tool" used to create the content.

  - **Users who prompt the models**: Some argue that the user who inputs the prompts or commands should hold the IP rights because they are directing the creation of the content.

  - **Owners of training data**: There could be claims from the entities that own the datasets the AI was trained on, especially if the output closely mirrors the input data.

  - **Commissioning parties**: In cases where AI is created for a specific purpose by a commissioning party, the contract terms may specify that this party owns the IP.

- **Data as IP**:

  - **Ownership of data**: Data used to train AI models can be viewed as valuable IP. Companies and institutions that contribute data may have IP claims, especially when the output generated closely mirrors the input data.

  - **Protection and usage**: It's critical to ensure data is used according to legal and contractual agreements, maintaining the integrity of data as IP during AI training and deployment processes.

- **Evolving legal frameworks**:

  - **Adapting laws**: As AI becomes more prevalent, there is a significant push to adapt IP laws to better define how AI-generated content is treated.

  - **Jurisdictional differences**: Different countries have different IP laws, leading to varying interpretations of who owns AI-generated content. For instance, the European Union has considered granting a form of copyright to the creators of AI systems, while other jurisdictions remain more traditional in their approach.

- **AI in IP enforcement**:

  - **Automated enforcement**: AI technologies can be leveraged to automatically detect IP infringements, such as unauthorized usage of copyrighted materials. AI can scan vast amounts of content to identify potential IP violations, providing an efficient tool for enforcement.

  - **Monitoring and alerts**: AI systems can continuously monitor the internet and digital spaces for instances of IP infringement, triggering alerts and initiating legal actions when necessary.

- **Ongoing debate and considerations**:

  - **Economic rights**: Who benefits economically from AI-generated content? Is it the developers, the users, or another party?

- **Moral rights**: Typically, copyright law includes moral rights, such as the right of attribution and the right to object to derogatory treatment of the work. How do these apply when the "author" is AI?

- **Liability and enforcement**: If AI-generated content infringes on existing copyrights, who is liable? Additionally, how are IP rights enforced in the digital realm where content can be easily and rapidly disseminated?

- **AI and trade secrets**:

  - **Protection of confidential information**: AI models may inadvertently expose sensitive information or trade secrets if improperly handled. Careful attention to how models are trained and how outputs are shared is critical to preventing unauthorized disclosure of proprietary information.

  - **Securing trade secrets**: Ensuring that trade secrets are not compromised during AI training or by model outputs requires strict confidentiality and secure data handling throughout the process.

- **The future of AI and IP**:

  - **Legislative action**: Some governments are beginning to explore legislation that would address the unique challenges posed by AI and IP rights. This includes considering whether AI can be a copyright holder or if new categories of protection are needed.

  - **Industry standards**: Organizations and corporations are also developing their own standards and practices for dealing with IP in AI-generated content, which could influence future laws and regulations.

Liability issues and LLM outputs

Liability issues related to the outputs of LLMs are a critical aspect of the legal and ethical framework within which these technologies operate. These concerns can have far-reaching implications for developers, companies, and users alike.

- **Liability and legal consequences**:

  - **Incorrect information**: If an LLM provides incorrect information that leads to financial loss, damage to reputation, or other harms, the question arises as to who is legally responsible for these consequences.

  - **Harmful content**: There is a risk that an LLM might generate content that is harmful, such as hate speech or libel, which could have legal ramifications.

  - **Legally sensitive information**: LLMs could inadvertently produce content that is legally sensitive, such as personal data that should be kept confidential, potentially violating privacy laws.

- **Responsibility and accountability**:

    - **Developers and companies**: Generally, the creators and distributors of LLMs may be held liable for their outputs. This potential for liability can extend to those who deploy LLMs in their applications or services.

    - **User agreements**: To mitigate liability risks, companies often include disclaimers and terms of service that limit their responsibility for the outputs of their LLMs.

    - **Regulations**: There is an increasing call for clear regulations that delineate the extent of liability for AI outputs. These regulations could help to establish standards for accountability and remedy.

- **Mitigating liability**:

    - **Disclaimers**: Companies typically use disclaimers to inform users that the outputs from LLMs are generated by algorithms and may not always be accurate or appropriate.

    - **User agreements**: These agreements can specify the acceptable use of an LLM and disclaim responsibility for misuse or reliance on the LLM's outputs.

    - **Transparency**: Providing transparency about the capabilities and limitations of LLMs can help set realistic expectations for users and may reduce legal risks.

- **Rigorous testing and validation**:

    - **Quality assurance**: Before deployment, LLMs must undergo rigorous testing to ensure that they function as intended and to minimize the risk of harmful outputs.

    - **Validation processes**: Continuous validation processes are essential to ensure that the LLM remains reliable and adheres to legal and ethical standards.

    - **Monitoring**: Post-deployment monitoring is crucial to quickly identify and rectify any issues that could lead to liability.

- **Ethical considerations**:

    - **Ethical guidelines**: Adhering to ethical guidelines in the development and deployment of LLMs can reduce the risk of outputs that could lead to legal issues.

    - **Human oversight**: Incorporating human oversight in the use of LLMs can help prevent problematic outputs and provide a mechanism for accountability.

These legal challenges require a collaborative effort between legal experts, technologists, policymakers, and ethicists to develop comprehensive guidelines and regulations that can keep pace with AI's rapid advancement. It is essential to establish clear legal principles that can guide the responsible deployment of LLMs while fostering innovation and protecting the rights and safety of individuals and organizations.

Regulatory landscape and compliance for LLMs

The regulatory landscape for LLMs is a complex and rapidly changing field, which organizations must carefully navigate to ensure compliance and avoid legal pitfalls. Here is a detailed examination of the current state and considerations:

- **Evolving regulatory environment**:

  - As AI technology advances, so does the regulatory framework that governs its use. Organizations using LLMs must stay abreast of both global and local regulations that could impact various aspects of LLM deployment.

  - This includes understanding restrictions on data usage, requirements for transparency in AI decision-making processes, and mandates for human oversight in critical applications.

- **Diverse requirements for AI systems**:

  - Different regions and countries may have varying requirements and standards for AI systems. For instance, the European Union's **General Data Protection Regulation (GDPR)** imposes strict rules on data privacy and users' rights to explanations for automated decisions, which directly affect how LLMs can be utilized.

  - In the United States, there may be sector-specific guidelines to consider, such as those pertaining to healthcare or financial services, which could influence the deployment of LLMs in these sectors.

- **Compliance with GDPR and other regulations**:

  - GDPR, in particular, has set a precedent for data protection laws worldwide. It requires that organizations protect the personal data and privacy of EU citizens for transactions that occur within EU member states. For LLMs, this means ensuring that any personal data used for training or output generation is handled according to GDPR stipulations.

  - GDPR also provides for the right to explanation, meaning that users have the right to understand the workings and decisions of algorithms affecting them, which requires LLMs to have a level of interpretability.

- **Awareness of AI-specific future legislation**:

  - It's not enough to comply with current regulations; organizations must also anticipate future changes in the legal landscape. This includes tracking proposals and discussions around AI-specific legislation, which could introduce new compliance requirements or restrictions.

  - Being proactive in these areas can help organizations adapt more readily to legal changes, ensuring continuous compliance and minimizing disruption to their operations.

- **Risk assessment and management**:

  - Conducting regular risk assessments regarding the use of LLMs can help identify areas where regulatory compliance may be at risk. This includes the evaluation of data sources, processing activities, and the potential impact of LLM outputs on users.

  - Developing a risk management strategy that includes plans for adapting to new regulations can help mitigate potential compliance issues before they arise.

In summary, as LLM use grows, a robust, proactive approach to regulatory compliance is crucial. Organizations must monitor legal developments, understand their impact, and adapt practices to meet regulatory requirements, including user data protection, transparency, and future legislative changes.

Ethical considerations and future outlook

The ethical deployment and use of LLMs are paramount to ensuring that these powerful tools benefit society without causing unintentional harm. Here's a deeper examination of the ethical considerations and what the future may hold in this space.

Transparency

Transparency in the context of LLMs is a foundational principle that serves multiple purposes, from fostering trust to ensuring accountability and enabling informed usage. A detailed exploration of why transparency is essential and what it entails is as follows:

- **Building trust with users and stakeholders**:

  - **Understanding model capabilities**: Clear communication about what LLMs can and cannot do helps set realistic expectations. Users need to be aware of the model's strengths, such as language understanding and generation, and its limitations, such as lack of real-world awareness or common sense.

  - **Data training disclosure**: Disclosure of the nature and source of the data LLMs are trained on is important for users to understand potential biases or the context in which the model performs best. For instance, if a model is trained predominantly on English internet text, its understanding of cultural nuances in other languages may be limited.

  - **Error and limitations acknowledgment**: LLMs, like any other AI system, are not infallible. They can make mistakes or produce unexpected results. Transparency about these limitations can help users make better-informed decisions about how to use and when to rely on the model's outputs.

- **Openness about methodologies and algorithms:**

  - **Scrutiny and improvement**: When the methodologies and algorithms used in LLMs are open to the public, it allows for academic and peer review, which can lead to improvements in the models. This collaborative approach can help to identify errors, reduce bias, and develop best practices.

  - **Replicability**: Transparency in AI is linked to the scientific principle of replicability. If other researchers or developers can understand and replicate the results of an LLM, this contributes to the robustness and credibility of the technology.

  - **Ethical considerations**: Openness about algorithms can also allow for ethical analysis and ensure that AI development aligns with societal values and norms. This is particularly important as LLMs become more integrated into critical aspects of society and individual daily lives.

- **Impact on end users and affected parties:**

  - **Informed consent**: Users should have the information necessary to provide informed consent when they interact with LLMs, especially when personal data is involved.

  - **Impact awareness**: Understanding how LLMs work is also important for those indirectly affected by their applications, such as people subject to decisions made with the assistance of LLMs in areas such as hiring, lending, or legal judgments.

- **Regulatory compliance:**

  - **Adhering to laws**: As mentioned earlier, various jurisdictions enact regulations that require transparency in AI systems. For example, GDPR has provisions for the right to explanation when automated decision-making is involved.

  - **Standardization**: Transparency helps in creating standards for AI systems that can facilitate compliance with such regulations across different regions and industries.

- **Challenges to transparency:**

  - **IP**: While openness is important, it must be balanced against the protection of IP since the development of LLMs involves significant investment and innovation.

  - **Complexity**: The complexity of LLMs can make transparency challenging. It can be difficult to explain intricate algorithms and data processing methods in a way that is accessible to non-experts.

  - **Security**: There is also a need to consider security implications, as revealing too much about the inner workings of an LLM could potentially expose vulnerabilities.

Accountability

Accountability in the deployment of LLMs is a critical aspect of their governance and operational integrity. It involves establishing responsibility for the actions of the models and ensuring that there are systems in place to correct any negative outcomes. Let's go through a detailed discussion of accountability in the context of LLMs.

- **Defining lines of accountability**:

    - **Responsibility assignment**: It is essential to determine who is responsible for the various aspects of an LLM's operation. This could include the developers, the organization deploying the LLM, the end users, or a combination thereof.

    - **Legal and ethical standards**: Accountability must be aligned with both legal requirements and ethical standards. It ensures that the use of LLMs complies with societal norms and regulations, such as data protection laws and non-discrimination principles.

- **Protocols for addressing issues**:

    - **Incident response plans**: Organizations must have plans to quickly and effectively respond to issues such as the spread of false information or the perpetuation of harmful biases.

    - **Monitoring systems**: Continuous monitoring can help detect when LLMs generate inappropriate or harmful content. This can include both automated systems and human oversight.

    - **Feedback loops**: There should be mechanisms for users to report problems and for those reports to be addressed. This feedback is crucial for improving the model and its governance.

- **Mechanisms for corrective action**:

    - **Human intervention**: The ability of humans to intervene in automated processes is a key aspect of accountability. If an LLM's output is questionable or problematic, human judgment should be applied to correct the issue.

    - **Audit trails**: Keeping records of the LLM's activity can help trace the cause of any issues and is essential for auditing and improving the system.

    - **Updating procedures**: When an issue is identified, there must be procedures in place to update the LLM, whether through retraining, tweaking the algorithm, or adjusting the input data.

- **Transparency and accountability**:

    - **Clear communication**: Part of being accountable is being transparent about how LLMs work, their limitations, and the steps being taken to mitigate risks.

    - **Documentation**: Comprehensive documentation of design choices, training data, and operational protocols supports accountability by providing a clear record that can be reviewed and assessed.

- **Ethical considerations**:

    - **Bias mitigation**: Ethical accountability includes the commitment to identify and reduce biases in LLMs. This might involve diversifying training data or developing algorithms that can detect and correct biases.

    - **Fairness and non-discrimination**: Ensuring that LLMs treat all users and groups fairly is a crucial part of accountability. This may involve ethical reviews and adherence to fairness protocols.

- **Accountability in practice**:

    - **Regulatory compliance**: Organizations must comply with any regulations that govern the use of AI and LLMs, such as the GDPR in Europe or the **California Consumer Privacy Act (CCPA)** in the United States.

    - **Industry standards**: Following industry standards and best practices can also help establish and maintain accountability.

Future outlook

The future outlook for AI, particularly in the context of its ethical considerations, presents a landscape where the pace of technological advancement is matched by a parallel development of ethical frameworks and review processes. Here's a comprehensive exploration of what this future might entail:

- **Continuous ethical assessments**:

    - **Dynamic ethical standards**: As AI technology evolves, so must the ethical standards that govern it. This is not a static field; what is considered ethical today may change as society evolves and new implications of AI are discovered.

    - **Ethical guidelines development**: Continuous ethical assessments will become integral to AI research and development, requiring AI practitioners to stay informed about current ethical guidelines and best practices.

    - **Real-time ethical decision-making**: AI systems might need to incorporate mechanisms for real-time ethical decision-making, especially in scenarios where AI actions have immediate consequences on individuals or society.

- **Integration of ethical reviews**:

    - **Standardization of ethical reviews**: Ethical reviews could become standardized across the AI industry, drawing parallels from established fields such as healthcare, where ethical review boards are a norm.

    - **Ethical certification**: Similar to how buildings have safety certifications, AI applications may have ethical certifications indicating that they have passed certain ethical standards and reviews.

- **Cross-disciplinary teams**: AI development teams might regularly include ethicists, sociologists, and legal experts to provide diverse perspectives on the potential impacts of AI.

- **Societal values and norms alignment**:

  - **Cultural sensitivity**: AI systems will need to be sensitive to a variety of cultural norms and values. This requires a global perspective on ethics, as AI technologies often cross geographical and cultural boundaries.

  - **Public participation**: There may be increased public participation in the ethical review process, with stakeholders from various sectors of society contributing to the discussion on AI ethics.

  - **Ethics in AI education**: Educational curricula for AI professionals are likely to include a strong component of ethics training, preparing the next generation of AI developers to think critically about the ethical implications of their work.

- **Evolving legal frameworks**:

  - **Regulatory response**: As ethical considerations gain prominence, regulatory frameworks around AI will likely evolve to incorporate ethical guidelines into legal requirements.

  - **International cooperation**: Given the global nature of AI, there may be increased international cooperation to develop and harmonize ethical standards across borders.

- **Proactive ethical design**:

  - **Ethics by design**: AI systems will be designed with ethical considerations in mind from the outset, rather than as an afterthought. This "ethics by design" approach will be fundamental to AI development practices.

  - **Preventative ethics**: The emphasis will shift toward preventative ethics—anticipating and designing out ethical risks before they materialize, rather than reacting to ethical lapses after they occur.

- **Technological considerations**:

  - **Transparency and explainability**: There will be a continued push for greater transparency and explainability in AI systems, allowing for ethical scrutiny and trust-building with users.

  - **Human-centric AI**: AI development will focus on human-centric principles, ensuring that AI serves to augment human abilities and improve well-being without infringing on individual rights or autonomy.

Continuous ethical assessments

Continuous ethical assessments in the context of LLMs are a vital component of responsible AI development and deployment. They involve ongoing evaluation and reflection on the ethical implications of these technologies. Here's a more detailed look at what continuous ethical assessments might entail:

- **Regular ethical evaluations**:

  - **Periodic review cycles**: Just like software undergoes regular updates and maintenance, ethical assessments of LLMs will require periodic reviews. These reviews would evaluate recent advancements, integration into new applications, and any societal shifts that might influence ethical perspectives.

  - **Adaptive ethical frameworks**: As technology evolves, so must the frameworks that assess its ethical use. Ethical guidelines will need to be dynamic, with the capacity to adapt to new developments in AI capabilities.

- **Multidisciplinary committees**:

  - **Diverse expertise**: Ethical assessments can benefit from the insights of a multidisciplinary committee that includes ethicists, technologists, sociologists, legal experts, and representatives from the public.

  - **Stakeholder engagement**: Including a broad range of stakeholders ensures that multiple perspectives are considered, especially those of groups that may be disproportionately affected by LLMs.

- **Ethical AI frameworks and toolkits**:

  - **Guidance tools**: Frameworks and toolkits can provide structured guidance to developers, helping them to consider the ethical implications of their work at each stage of the development process.

  - **Best practices and standards**: These tools can also help establish industry-wide best practices and standards for ethical AI development.

- **Contextual considerations**:

  - **Context-specific assessments**: The impact of LLMs can vary greatly depending on the context in which they are used. Ethical assessments must take into account the specific use cases, from healthcare to finance to education.

  - **Cultural sensitivity**: Global deployment of LLMs requires sensitivity to different cultural norms and values. Ethical assessments will need to consider the diversity of global users and stakeholders.

- **Impact evaluation**:

  - **Direct and indirect effects**: Evaluations must consider both the direct effects of LLM outputs and the indirect effects, such as the impact on employment or societal trust.

  - **Long-term implications**: Ethical assessments should also consider the long-term societal implications of LLM integration, including potential shifts in power dynamics or information control.

- **Proactive measures**:

  - **Anticipatory ethics**: Instead of being reactive, ethical assessments should anticipate potential ethical issues and address them proactively.

  - **Ethics in design**: Incorporating ethical considerations from the very beginning of the design process, known as "value-sensitive design," can help to embed ethical principles into the technology itself.

- **Scalability and evolution**:

  - **Scalable processes**: As LLMs become more widely used, the processes for ethical assessment will need to be scalable to keep up with the pace of AI deployment.

  - **Evolving guidelines**: Ethical guidelines will evolve as more is learned about the capabilities and impacts of LLMs, and as societal values themselves change over time.

In conclusion, the ethical considerations surrounding LLMs demand a proactive and ongoing commitment to transparency and accountability. As we look to the future, continuous ethical assessments and the integration of ethical considerations into the AI development lifecycle will be critical for guiding the responsible advancement of this technology. Ensuring that LLMs are used ethically will require collaboration across sectors and disciplines and a shared commitment to prioritizing the well-being of individuals and society.

Hypothetical case study – bias mitigation in AI for hiring platforms

In 2023, a large tech company launched an AI-powered hiring tool designed to streamline the recruitment process by analyzing resumes and recommending the best candidates. The tool, based on machine learning algorithms and an LLM, was trained on historical data of past hiring decisions made by the company.

Initial issue

Despite its advanced capabilities, the AI system began to exhibit significant gender biases. It favored male candidates over female ones for technical positions, reflecting the historical bias embedded in the company's prior hiring data. The model learned patterns that perpetuated gender imbalances rather than mitigating them. This bias raised ethical, legal, and operational concerns, putting the company at risk of discrimination lawsuits and reputational damage.

Bias mitigation approach

To address this issue, the company implemented a multi-step bias mitigation strategy:

1. **Dataset curation**: The development team revisited the training data and identified the biased patterns present. They removed gender-specific indicators from the data, such as gendered pronouns and references, and ensured the data was more representative of diverse candidate backgrounds.

2. **Secure data handling**: In order to prevent sensitive candidate information from being misused or exposed, the company enforced strict access controls and anonymized the dataset. This anonymization process also helped in reducing bias by removing irrelevant personal identifiers that could influence hiring decisions, such as gender or age.

3. **Algorithmic auditing**: The system underwent continuous auditing, using fairness metrics to assess whether its recommendations exhibited any form of bias. The AI model was also subjected to adversarial tests to ensure it could handle inputs from a diverse pool of candidates without reverting to biased patterns.

4. **Human oversight and explainability**: The company introduced human oversight to review the AI's final recommendations. The development team implemented explainability features, allowing hiring managers to understand why the model recommended specific candidates and ensure that the AI's decision-making was transparent.

5. **AI in IP enforcement**: As the system was further refined, the company integrated AI-based IP enforcement to protect proprietary algorithms. Automated IP enforcement tools were employed to detect unauthorized usage or reproduction of their AI hiring platform, safeguarding their innovations while maintaining the integrity of the revised, bias-mitigated model.

Outcome

After implementing these measures, the bias in the hiring process was significantly reduced. The AI system began recommending a more diverse group of candidates, improving gender representation in the company's technical teams. Furthermore, with the incorporation of secure data handling practices, the company not only enhanced its ethical standing but also ensured compliance with privacy regulations such as GDPR.

Key takeaways

Here are the key takeaways from this case study:

- **Bias mitigation is essential**: This case demonstrates the practical importance of addressing bias in AI, particularly in systems that impact people's lives, such as hiring platforms

- **Continuous monitoring**: Ongoing evaluation of the model's performance and bias mitigation efforts ensured that the AI system did not revert to biased behaviors

- **Legal and ethical considerations**: Bias mitigation not only improves fairness but also shields organizations from legal risks, such as discrimination claims

- **Collaborative approach**: Engaging diverse stakeholders, including legal experts, AI developers, and HR teams, was crucial for refining the system to promote fairness and transparency

This case highlights the practical necessity of bias mitigation in LLMs, especially when these models are deployed in critical applications such as hiring. It demonstrates that addressing bias is not only a technical challenge but also a vital legal and ethical responsibility.

Summary

Securing LLMs is an essential and ongoing process that requires vigilance and a multi-layered strategy to counteract a spectrum of vulnerabilities. Adversarial attacks that manipulate data to deceive models must be countered with rigorous testing and well-crafted defenses. Regular vulnerability scanning and testing are crucial to uncover emerging threats, while proactive security measures and continuous security monitoring ensure that protections evolve in tandem with new attack vectors. Collaboration among developers, security experts, and the wider community enhances these efforts, forming a comprehensive defense against the misuse or malfunction of LLMs. These security practices, accompanied by continuous ethical assessments and updates, are integral to maintaining the integrity, performance, and reliability of LLMs, thereby ensuring they are aligned with evolving societal values and legal standards.

In the next chapter, we present case studies with business applications and a discussion on **return on investment (ROI)**.

12

Case Studies – Business Applications and ROI

In this chapter, we will examine the application and **return on investment** (**ROI**) of LLMs in business. We will start with their role in enhancing customer service, showcasing examples of improved efficiency and interaction. Our focus will then shift to marketing, exploring how LLMs optimize strategies and content. The next section will cover LLMs in operational efficiency, particularly in automation and data analysis. We will conclude by assessing the ROI from LLM implementations, considering both financial and operational benefits. Through these sections, we will present a comprehensive overview of LLMs' practical business uses and their measurable impacts.

In this chapter, we're going to cover the following main topics:

- Implementing LLMs in customer service enhancement
- LLMs in marketing – strategy and content optimization
- Operational efficiency through LLMs – automation and analysis
- Assessing ROI – financial and operational impacts of LLMs

By the end of this chapter, you should have gained a thorough understanding of the practical applications and ROI of LLMs in various business contexts.

Implementing LLMs in customer service enhancement

All three case studies in this chapter are created mock cases, engineered to provide relevant information.

Background

A leading telecommunications company, Comet Communications, faced challenges in managing the increasing volume of customer inquiries and support requests. Their traditional customer service infrastructure was strained, leading to long wait times and decreased customer satisfaction.

Objective

The objective was to improve customer satisfaction by providing quick, accurate, and personalized support while reducing the workload on human customer service agents.

Implementation of LLMs

The steps to fulfill our objective are discussed next.

Integration with existing infrastructure

The implementation of LLMs within existing customer service infrastructures, such as the case with Comet Communications, involves several strategic steps to enhance customer interaction and streamline service processes. Here's a detailed exploration of how such integration might occur:

1. **Assessing compatibility**: Before integration, a thorough assessment of the existing customer service infrastructure is necessary to determine compatibility with the LLM. This includes reviewing the current software platforms, databases, and customer interaction channels to ensure the LLM can be seamlessly incorporated without disrupting service.

2. **Designing the LLM for natural interactions**: The LLM is intricately designed to understand and process natural language input from customers. This involves the following:

   - **Natural language understanding** (**NLU**): Implementing NLU capabilities to accurately interpret customer queries, which may include slang, typos, and colloquial language

   - **Natural language generation** (**NLG**): Utilizing NLG for the LLM to craft responses that are coherent, contextually appropriate, and indistinguishable from human communication

3. **Customizing responses**: Customization involves tailoring the LLM's responses to fit the brand's voice and policies. It also means ensuring that responses are personalized based on the customer's history and preferences, which requires integration with **customer relationship management** (**CRM**) systems.

4. **Data integration**: Integrating the LLM with the company's databases allows it to access relevant information, such as product details, service status, and customer account information. This enables the LLM to provide informed and accurate responses to specific inquiries.

5. **Multichannel deployment**: The LLM is deployed across various customer service channels, including the following:

   - **Live chat**: Assisting customers in real time on the company website or mobile app

   - **Email**: Automatically generating responses to customer emails with the option for escalation to human agents for complex issues

   - **Social media**: Monitoring and responding to customer queries on platforms such as Twitter, Facebook, and Instagram

6. **Continuous learning and improvement**: The LLM is set up for continuous learning, allowing it to improve over time in the following ways:

   - **Feedback loops**: Incorporating customer feedback and agent reviews to enhance the accuracy and helpfulness of the LLM's responses

   - **Machine learning**: Using machine learning algorithms to refine the model's predictions and responses based on new data

7. **Ensuring compliance and security**: Ensuring that the LLM adheres to privacy laws and security standards is paramount. This involves the following:

   - **Data privacy**: Implementing protocols for data anonymization and encryption to protect customer information

   - **Compliance checks**: Regularly reviewing interactions for compliance with industry regulations and company policies

8. **Performance monitoring**: The LLM's performance is constantly monitored using metrics such as response accuracy, customer satisfaction scores, and resolution times to ensure it meets performance standards.

9. **Human oversight**: Establishing a system of human oversight where customer service agents can intervene when necessary, either for complex issues the LLM cannot handle or for quality assurance purposes.

10. **Scaling and evolution**: As the LLM proves effective, plans for scaling its use are developed, including expanding language capabilities, enhancing feature sets, and integrating with additional business processes and tools.

Training the LLM

Training an LLM for customer service applications is a critical step that determines the effectiveness of the AI in handling real-world customer interactions. The process typically involves the following detailed steps:

- **Data collection**:

  - **Historical interactions**: The initial step involves gathering extensive historical data, which includes previous customer service interactions across various channels. This data serves as the foundation for the LLM's learning.

  - **Data diversity**: The dataset should encompass a wide range of scenarios, from simple FAQs to complex problem-solving interactions, to provide the LLM with a comprehensive understanding of possible customer service situations.

- **Data preparation**:

  - **Cleaning and anonymization**: Data must be cleaned to remove any irrelevant or redundant information and anonymized to protect customer privacy.

  - **Labeling**: Relevant data may need to be labeled to facilitate supervised learning. For instance, customer questions can be tagged with the appropriate category or intent.

- **Specialized training**:

  - **Company-specific jargon**: It's crucial for the LLM to understand and use the specific terminology associated with the company's products and services. This often involves creating a glossary or lexicon for the LLM to learn from.

  - **Common issues**: The model is particularly trained on the most frequent customer issues identified from the data. This ensures that the LLM can handle these high-volume queries with a high degree of accuracy.

- **Model development**:

  - **Algorithm selection**: Choosing the right machine learning algorithms is key to developing an effective LLM. This might involve decisions between different neural network architectures such as transformers or recurrent neural networks.

  - **Feature engineering**: Identifying and engineering the right features from the data can help the LLM better understand the context and content of customer interactions.

- **Iterative training and validation**:

  - **Initial training**: The LLM is first trained (either fine-tuned from an existing pre-trained LLM or trained from scratch) on a subset of the data to learn basic patterns and responses.

  - **Validation and testing**: Regular cycles of validation and testing with separate datasets are conducted to assess the model's performance and make necessary adjustments.

- **Evaluation metrics**:

  - **Accuracy**: The LLM's ability to correctly understand and respond to queries is measured using accuracy metrics.

  - **Response quality**: Beyond accuracy, the quality of the responses is evaluated, often through human review or customer satisfaction surveys.

- **Continuous learning**:

  - **Real-time learning**: Once deployed, the LLM continues to learn from ongoing customer interactions, allowing it to adapt to new trends or changes in customer behavior.

- **Feedback incorporation**: Customer service agents and customer feedback play a role in the LLM's ongoing training, highlighting areas for improvement or additional training needs.

- **Monitoring and updating**:

  - **Performance monitoring**: The LLM is continuously monitored for performance dips or issues, which could indicate the need for retraining or updates.

  - **Model refinement**: Based on performance data and feedback, the LLM is periodically refined and updated to improve its interactions and responses.

After implementing the previous steps, the following actions took place:

- **Pilot program**:

  - Comet launched a pilot program, routing a portion of customer inquiries through the LLM to gauge effectiveness

  - The LLM was configured to escalate more complex issues to human agents

- **Feedback loop**:

  - Customers were surveyed post-interaction to measure satisfaction levels

  - Customer service agents provided feedback on the LLM's performance, which was used for iterative improvements

- **Ethical considerations**:

  - Transparency measures were put in place, informing customers that they were interacting with an AI

  - Anonymized data was used to train the model to respect customer privacy

Results

The following positive results took place:

- **Increased efficiency**:

  - The LLM handled a significant volume of routine inquiries, reducing the average handling time per query

  - Human customer service agents were able to focus on complex and high-priority cases

- **Customer satisfaction**:

  - Post-implementation surveys indicated an improvement in customer satisfaction due to quicker resolution times and 24/7 availability of support

- Personalized support, with the LLM pulling up customer history and preferences, led to more targeted and helpful interactions

- **Cost savings**:

  - Comet Communications reported a reduction in operational costs associated with customer support

  - The need for a large-scale human customer service team was reduced, allowing for resource reallocation to other strategic areas

Challenges

There were a few challenges:

- **Continual training**:

  - The LLM required ongoing training to keep up with new products, services, and emerging customer service issues

  - Regular updates were necessary to maintain the model's accuracy and relevance

- **Human-AI collaboration**:

  - Ensuring a smooth handover from the LLM to human agents in complex cases required fine-tuning the escalation protocols

  - Training for human agents was necessary to effectively collaborate with the LLM and leverage its capabilities

Future developments

The following actions took place post-implementation:

- **Advanced personalization**: Future updates to the LLM aimed to include more advanced personalization, using machine learning to predict customer needs and provide proactive support

- **Multilingual support**: Plans were in place to train the LLM in multiple languages, catering to a diverse customer base and expanding global reach

- **Wider implementation**: Based on the success of the pilot, Comet Communications planned to implement the LLM across all customer service channels, including voice support

Conclusion

The case of Comet Communications demonstrates the potential of LLMs to transform customer service by enhancing efficiency, personalization, and customer satisfaction while also presenting opportunities for cost savings and business growth. Continual learning and adaptation are key to maintaining the effectiveness of such AI applications in dynamic business environments.

LLMs in marketing – strategy and content optimization

The following case is about a company implementing LLMs for their marketing strategy and content optimization.

Background

In the fast-paced world of digital marketing, Digimarket Corporation, a multinational consumer goods company, recognized the need to optimize its marketing strategy and content creation to stay ahead of the competition. The company sought to leverage the capabilities of LLMs to enhance its marketing efforts.

Objective

Digimarket Corporation aimed to use LLMs to generate engaging content, personalize customer interactions, and analyze market trends to improve its overall marketing strategy.

Implementation of LLMs

Let's go through the LLM implementation actions that were taken.

Content creation and personalization

The implementation of LLMs for content creation and personalization is a transformative approach that businesses such as Digimarket Corporation can utilize to streamline their marketing efforts. Here is how the implementation process typically unfolds and the benefits it brings:

- **Planning and strategy development**:

    - **Brand consistency**: Before implementation, Digimarket Corporation would establish guidelines to ensure that the LLM-generated content aligns with the brand's voice and messaging strategy.

    - **Content types**: The company decides which types of content the LLM will generate. This could range from short-form social media updates to long-form blog posts and detailed product descriptions.

- **LLM training for content generation**:

    - **Dataset assembly**: The LLM is trained on a comprehensive dataset of the company's past marketing materials, including successful ad copy, product descriptions, and blog posts, to learn the style and substance that resonates with the company's audience.

- **Customization for brand voice**: The training also includes company-specific jargon, slogans, and value propositions to maintain brand consistency in the generated content.

- **Personalization variables**: Training incorporates customer data to enable personalization, such as browsing history, purchase patterns, and demographic information.

- **Content creation workflow**:

  - **Automated content generation**: Once trained, the LLM auto-generates content at scale, which is then reviewed and fine-tuned by marketing teams to ensure it meets quality standards.

  - **Personalization engine**: The LLM personalizes content for different audience segments, adjusting the language, tone, and messaging based on the target demographic's preferences.

- **Review and optimization cycle**:

  - **Content review process**: A system is established for human editors to review and approve content, ensuring it aligns with brand guidelines and marketing objectives.

  - **Performance analysis**: The performance of LLM-generated content is continuously monitored using **key performance indicators** (**KPIs**) such as engagement rates, **Click-Through Rates** (**CTRs**), and conversion rates.

  - **Feedback loop**: Data from content performance feeds back into the LLM to refine and improve future content generation.

- **Integration with marketing campaigns**:

  - **Campaign coordination**: The LLM-generated content is integrated into various marketing campaigns, from email marketing to social media, ensuring a consistent and personalized customer experience.

  - **Dynamic content adjustment**: The LLM adapts content in real time based on campaign performance, audience interaction, and A/B testing results.

- **Benefits and outcomes**:

  - **Efficiency**: Digimarket Corporation benefits from increased efficiency in content production, with the LLM generating large volumes of content quickly.

  - **Scalability**: LLMs enable the company to scale content creation to meet the demands of various marketing channels without a proportional increase in resources.

  - **Engagement**: Personalized content has a higher likelihood of engaging customers, leading to better campaign performance and customer retention.

  - **Brand reinforcement**: Consistent brand messaging across all content reinforces brand recognition and loyalty.

- **Challenges and considerations**:

  - **Quality control**: Maintaining high-quality standards requires robust review processes and human oversight.

  - **Ethical use of data**: Personalization must be balanced with respect for customer privacy and compliance with data protection regulations.

SEO optimization for Digimarket Corporation (existing marketing case study)

Optimizing content to achieve higher rankings in **Search Engine Results Pages** (**SERPs**) and increase a website's organic traffic is a critical component of digital marketing known as **Search Engine Optimization** (**SEO**). When LLMs are trained for SEO optimization, several nuanced steps and considerations are involved to ensure the content meets the standards that search engines favor. Here's how LLMs can be leveraged for SEO optimization.

Training LLMs for SEO

- **Understanding SEO principles**:

  - The LLM is trained to understand key SEO principles, such as keyword relevance, content originality, and the importance of headings and meta tags.

  - Training includes learning from SEO-optimized content that has historically ranked well in SERPs.

- **Incorporating market trends**:

  - LLMs are provided with current market research and trend analyses to understand what content is most relevant and in demand.

  - This training helps the LLM identify and integrate trending topics and keywords into the content.

- **Algorithm updates**: SEO is a dynamic field with frequent search algorithm updates. The LLM must be updated with the latest search engine guidelines and ranking factors.

Content creation with SEO focus

- **Keyword optimization**:

  - The LLM uses natural language processing to incorporate relevant keywords into the content without keyword stuffing, ensuring readability while optimizing for search engines

  - It understands the importance of long-tail keywords and user intent, creating content that addresses what users are searching for

- **Content structuring**: Training includes structuring content with proper use of headers, subheaders, and bullet points, making it easier for search engines to crawl and index
- **Meta data generation**: The LLM is capable of generating meta titles and descriptions that are not only SEO-friendly but also compelling to users, thereby increasing CTRs from SERPs

Continuous learning and updating

- **Feedback loop**:

  - The performance of the content in SERPs is monitored, and the LLM is continuously updated with this feedback to improve future content

  - Adjustments are made based on which strategies lead to better rankings and which do not

- **Adaptation to SEO changes**: The LLM adapts to changes in SEO best practices and updates its content generation process accordingly

SEO content quality assurance

- **User Experience (UX) and engagement**: Beyond SEO, the LLM also learns to optimize content for UX, ensuring that visitors stay engaged, which is a factor in search engine rankings
- **Link-worthy content**: LLMs are trained to create content that is informative and authoritative, encouraging other websites to link to it, thereby improving backlink profiles

Measuring SEO success

- **Tracking organic traffic**:

  - Analytics tools are used to track increases in organic traffic that are attributable to the LLM-generated content.

  - Metrics such as bounce rate and session duration are also monitored to assess the engagement level of the content.

- **SERP position tracking**: The rankings of content pieces on SERPs are tracked to measure the effectiveness of the LLM's SEO strategies.

- **Keyword rankings**:

  - Monitoring keyword rankings helps assess how the content performs for targeted keywords over time.

  - Tools such as SEMrush or Ahrefs can be used to track fluctuations in rankings and identify which pages are improving or declining in visibility.

- **CTR**:

  - CTR measures the percentage of users who click on your content in search results.

  - A higher CTR indicates that the content title and meta descriptions are compelling, enticing users to visit the site.

- **Engagement metrics**: Engagement metrics, such as time on page and number of pages per session, provide insight into how users are interacting with the content. These metrics help gauge whether the content is holding the visitor's attention and encouraging deeper site exploration.

- **Content quality and relevance**:

  - Quality and relevance play a significant role in how search engines rank content.

  - Regular content audits ensure that the material is up-to-date, valuable to users, and in line with search intent.

  - High-quality content also improves engagement metrics and leads to better search rankings.

- **Conversion rate**:

  - Ultimately, the success of SEO efforts should be tied to conversions, whether that's purchases, sign-ups, or lead generation.

  - Conversion rate optimization ensures that the traffic driven by SEO is aligned with the website's business goals.

  - Monitoring conversion rates can highlight the effectiveness of the content in driving desired actions.

- **Backlinks**:

  - Backlinks remain a strong ranking factor in SEO. The quantity and quality of backlinks can significantly impact a site's authority and SERP rankings.

  - Tools such as Ahrefs or Majestic can be used to track the growth and quality of backlinks pointing to your content. This helps with identifying potential areas for link-building efforts.

- **Technical SEO metrics**:

  - Technical aspects of SEO, such as page load time, mobile-friendliness, and proper use of schema markup, can greatly affect rankings.

  - Tools such as Google Lighthouse and Screaming Frog help assess these technical factors.

- **Site speed**:

  - Page load times can impact both UX and rankings.

  - Faster sites tend to rank better and offer a more pleasant UX, reducing bounce rates.

- **Mobile optimization**: With Google's mobile-first indexing, ensuring your content is mobile-friendly is crucial for maintaining or improving SERP rankings.

- **Structured data (schema markup)**: Implementing structured data helps search engines better understand your content and can improve rankings by providing rich snippets in search results.

Social media strategy

LLMs analyzed social media trends and engagement patterns to optimize posting schedules and content types for Digimarket's social media accounts, increasing engagement rates and follower growth.

Campaign analysis

Post-campaign, the LLM analyzed consumer engagement data to evaluate the effectiveness of different content strategies and provided recommendations for future campaigns.

Email marketing

For email campaigns, the LLM generated subject lines and email content that led to higher open and CTRs, based on historical performance data.

Results

After implementing the previous steps, the following results were achieved:

- **Increased engagement**: Content generated by the LLM saw a significant increase in consumer engagement across various channels, with a noted improvement in user comments and shares on social media platforms

- **Higher conversion rates**: Personalized marketing messages resulted in higher conversion rates and an uptick in sales, particularly for targeted email campaigns

- **SEO success**: SEO-optimized content created by the LLM helped Digimarket Corporation achieve first-page rankings for several key product categories, leading to increased visibility and organic reach

- **Cost efficiency**: The use of LLMs reduced the overall cost of content production and strategy development, allowing marketing funds to be reallocated to other strategic initiatives

Challenges

The following challenges were encountered during the LLM implementation:

- **Content variety**: Ensuring a variety of content styles and avoiding repetitive phrasing required ongoing tuning of the LLM's parameters and creative oversight.

- **Brand voice consistency**: Maintaining a consistent brand voice across all generated content necessitated the development of comprehensive style guides for the LLM.

- **Algorithm bias**: Vigilance was needed to monitor for algorithmic bias that could lead to skewed content strategies or non-inclusive marketing practices. Information on algorithmic bias can be found in several other chapters in this book.

  Algorithmic bias occurs when repeatable and systematic errors in a computer system lead to unfair outcomes by favoring one arbitrary group of users over others. In the context of LLMs used for content creation and marketing strategies, this can manifest in several ways:

- **Sources of bias in LLMs**:

  - **Training data bias**: If the training data contains historical biases or underrepresentation of certain groups, the LLM may replicate or even amplify these biases in its output

  - **Selection bias**: This occurs when the data used to train the LLM is not representative of the diverse customer base or market segments

  - **Confirmation bias**: An LLM might generate content that aligns with popular or dominant views, potentially neglecting niche or contrarian perspectives, thereby reinforcing certain biases

- **Impacts of algorithmic bias**:

  - **Skewed content strategies**: Biased algorithms may lead to content that does not resonate with or even offends part of the target audience, resulting in ineffective marketing strategies

  - **Non-inclusive marketing practices**: If the content is biased, it may exclude or misrepresent certain demographic groups, leading to a lack of inclusivity in marketing campaigns

  Monitoring and mitigation of algorithmic bias can be achieved through the following practices:

  - **Diverse data sets**: Ensure that the training data is diverse and inclusive, representing a wide range of demographics, cultures, and languages

  - **Bias detection tools**: Utilize specialized tools and statistical methods to detect and measure bias in algorithm outputs

  - **Human oversight**: Employ human reviewers to assess and adjust the content before it goes live, ensuring that it aligns with ethical standards and is inclusive

  - **Regular audits**: Conduct regular audits of the LLM's performance to identify any biases, with a focus on outcomes and the decision-making process of the model

  - **Feedback mechanisms**: Implement feedback loops that allow for the reporting and addressing of bias, including input from diverse user groups

  - **Algorithmic transparency**: Strive for transparency in how the LLM operates, making it easier to identify the sources of any bias

  - **Ethical guidelines**: Develop and adhere to ethical guidelines for AI and data usage, which can guide the development and deployment of LLMs

Vigilance in monitoring for algorithmic bias is essential to prevent skewed content strategies and ensure marketing practices are inclusive. This involves a commitment to diversity in data, transparency in algorithms, and an ongoing effort to identify and mitigate biases. By addressing these issues proactively, Digimarket can build trust with its audience and foster a more equitable digital environment.

Future developments

The following actions occurred post-LLM implementation:

- **Interactive content**: Plans were underway to develop interactive marketing content using LLMs, enhancing customer experience through quizzes, personalized recommendations, and AI-driven interactive stories

- **Predictive analytics**: Digimarket Corporation intended to expand the use of LLMs in predictive analytics, forecasting market trends, and consumer behavior to stay ahead of market shifts

- **Omnichannel strategy**: The company aimed to implement an LLM-driven omnichannel strategy, providing a seamless customer experience across all digital and physical touchpoints
- **Future AI developments**: The company continuously keeps an eye out for future AI developments to be up-to-date

Conclusion

Digimarket Corporation's integration of LLMs into its marketing strategy yielded significant improvements in engagement, conversion rates, and cost efficiency. While challenges such as content variety and brand voice consistency remain, the potential for LLMs in marketing strategy and content optimization is vast and promises to reshape how companies interact with consumers. Continued innovation and vigilant management of these systems will be essential to harness their full potential while maintaining ethical marketing practices.

Operational efficiency through LLMs – automation and analysis

Let's go through a case study that exemplifies how a financial services provider implemented LLMs to achieve operational efficiency.

Background

TermCorp, a global financial services provider, faced challenges with operational efficiency, particularly in processing customer queries, generating reports, and analyzing financial documents. The company sought to automate these tasks and enhance its data analysis processes.

Objective

Their objective was to leverage LLMs in automating routine operations and analyzing large volumes of text-based financial data to improve efficiency, reduce errors, and expedite decision-making processes.

Implementation of LLMs

We will go through the steps that occurred during the implementation of LLMs.

Automated customer service

TermCorp implemented an LLM to handle first-level customer inquiries, automating responses to common questions and transactions, which reduced the workload on customer service representatives.

Document analysis and report generation

The LLM was trained to analyze financial documents such as balance sheets, earnings reports, and regulatory filings to extract relevant data points for analysis.

It was also used to automate the generation of financial reports, summarizing key metrics and insights in natural language that is easily understandable by stakeholders.

Data quality management

LLMs were utilized to scan through financial records, identifying inconsistencies and anomalies that could indicate errors or fraudulent activity, thus improving the accuracy of financial data.

Process optimization

By analyzing workflow data, the LLM identified bottlenecks and suggested process improvements, streamlining operations across various departments.

Process optimization through the use of LLMs is a cutting-edge approach to enhancing organizational efficiency.

Understanding process optimization

Process optimization involves analyzing existing workflows to identify inefficiencies, redundancies, or bottlenecks that can be restructured or eliminated to improve overall operational efficiency.

Role of LLMs in process optimization

The following explains what the role of LLMs is in process optimization:

- **Data analysis**:
  - LLMs can analyze large volumes of text-based workflow data, including process documentation, employee feedback, and performance reports
  - By processing natural language data, LLMs can understand the context and content of workflows, not just numerical data

- **Bottleneck identification**:
  - Using pattern recognition and anomaly detection, LLMs can identify irregularities and inconsistencies in processes that might indicate bottlenecks
  - They can analyze unstructured data, such as written reports and logs, to pinpoint areas where delays commonly occur

- **Suggesting improvements**:

  - LLMs can suggest process improvements by comparing current workflows to best practices learned during training

  - They can simulate different scenarios and predict the outcomes of proposed changes to the workflow

- **Cross-departmental analysis**:

  - LLMs can analyze data across various departments to ensure that process optimizations align with the overall organizational workflow and do not negatively impact other departments

Implementing LLM-driven process improvements

The following list represents how to implement LLM-driven process improvements:

- **Integrating with existing systems**: For LLMs to access relevant workflow data, they need to be integrated with the company's **enterprise resource planning** (ERP) systems, project management tools, and other operational software

- **Feedback loop**: Establish a feedback loop where employees can provide input on the LLM's suggestions, which can be used to refine the optimization models

- **Change management**: Implementing any process changes suggested by LLMs requires careful change management to ensure employee buy-in and minimal disruption to operations

Benefits of LLMs in process optimization

The benefits of LLMs in process optimization include the following:

- **Efficiency gains**: By automating the analysis of workflow data, LLMs can quickly identify areas for improvement, leading to faster and more efficient operations

- **Cost reduction**: Streamlined processes often result in reduced operational costs, as resources are used more effectively, and time is saved

- **Scalability**: LLMs can handle increasing volumes of data as the company grows, ensuring that process optimization is scalable and adaptable

- **Continuous improvement**: With continuous learning capabilities, LLMs can adapt to new data and changing conditions, promoting a culture of ongoing improvement

Challenges and considerations

The following are challenges and considerations regarding the implementation of LLMs in this case study:

- **Complexity of workflows**: Complex workflows may be challenging for LLMs to fully comprehend, requiring a combination of AI and human expertise

- **Data privacy and security**: Ensuring that sensitive process data is handled securely and in compliance with privacy regulations is crucial

- **Employee engagement**: Employees need to be engaged in the optimization process, with clear communication about the role of LLMs and the value they bring

Results

The results of this case study's LLM implementation are as follows:

- **Increased efficiency**:

    - Automation of customer service inquiries led to a 40% reduction in response times and a 30% decrease in the need for escalations to human representatives

    - Report generation time was reduced by 50%, with the added benefit of improved report consistency and clarity

- **Enhanced data accuracy**: The error rate in financial data processing was reduced significantly, resulting in higher trust from clients and stakeholders

- **Improved decision-making**: The insights derived from LLM analysis provided executives with actionable intelligence, leading to more informed decision-making

- **Cost savings**: TermCorp reported a noticeable decrease in operational costs due to reduced manual labor and fewer errors

Challenges

Here are the challenges that were encountered:

- **Integration with legacy systems**: Integrating LLMs with TermCorp's existing IT infrastructure required significant upfront investment and technical expertise

- **Continuous training and updating**: The LLM needed ongoing training to keep up with the latest financial regulations and company policies

- **User acceptance**: Encouraging employees to trust and effectively use the LLM was initially challenging, necessitating comprehensive training and change management initiatives

Future developments

The following actions and considerations occurred post-LLM implementation:

- **Predictive analytics**: Plans to enhance the LLM's capabilities with predictive analytics were put in place, aiming to forecast market trends and customer behaviors

- **Expansion to other operations**: TermCorp intended to expand the use of LLMs to other operational areas, such as risk management and compliance monitoring

- **Personalized customer insights**: By analyzing customer interactions and feedback, the LLM would provide personalized insights, enabling TermCorp to offer tailored financial advice and products

Conclusion

TermCorp's integration of LLMs into their operational processes led to enhanced efficiency, improved data accuracy, and cost savings. The successful implementation demonstrated the potential of LLMs to revolutionize operations in the financial industry. As the technology continues to evolve, TermCorp is positioned to further capitalize on LLMs for automation and analysis, setting a benchmark for operational excellence in the sector.

Assessing ROI – financial and operational impacts of LLMs

Assessing the ROI for implementing LLMs involves a comprehensive analysis of both the financial and operational impacts on an organization. Here's how organizations typically approach this assessment.

Financial impact assessment

In the realm of financial impact assessment, organizations meticulously calculate cost savings and weigh them against revenue growth opportunities. They consider initial investment costs and aim to determine the payback period. Simultaneously, operational impact assessment evaluates efficiency gains, quality of service improvements, scalability, and the role of LLMs in fostering innovation and competitiveness.

The following are the financial impacts to take into consideration when assessing ROI:

- **Cost savings**:

  - **Reduction in labor costs**: Organizations calculate the reduction in labor costs due to automation of tasks that were previously handled by employees

  - **Decrease in error-related expenses**: By minimizing errors in processes such as data entry, companies save on rectification costs

- **Revenue growth**:

  - **Enhanced sales**: By improving customer service and marketing, LLMs can lead to an increase in sales due to improved customer satisfaction and engagement

  - **New revenue streams**: The introduction of new, AI-driven products or services can open additional revenue channels

- **Investment costs**:

  - **Initial setup and integration**: The costs of implementing LLMs, including purchasing, developing, and integrating the technology into existing systems

  - **Training and development**: Investments in training for staff to use LLMs and ongoing model training to maintain performance

- **Payback period**: This involves determining how long it takes for the cost savings and additional revenue to cover the initial investment in the LLM technology

Operational impact assessment

Organizations are continually seeking ways to enhance their operational efficiency and service quality. An operational impact assessment provides a structured analysis of the direct effects that new technologies, such as LLMs, have on an organization's workflow, resource management, and overall service delivery. This assessment focuses on several key areas and by quantifying the tangible benefits that LLMs bring to these aspects, businesses can make informed decisions about the integration of these advanced tools into their operations, ensuring they remain competitive and responsive to market demands and customer needs.

The key areas that are relevant here include the following:

- **Efficiency gains**:

  - **Process automation**: Quantifying the time savings from automating routine tasks and the increased throughput of operational processes

  - **Resource allocation**: Assessing how LLMs free up employee time to focus on higher-value tasks

- **Quality of service**:

  - **Customer satisfaction**: Measuring changes in customer satisfaction metrics can indicate the impact of LLMs on service quality

  - **Accuracy of information**: LLMs often provide more accurate information handling, which can be quantified by reduced customer complaints or increased compliance rates

- **Scalability**:

  - **Handling volume**: Evaluating the LLM's ability to handle increased volumes of work without a corresponding increase in errors or delays

  - **Flexibility**: Assessing how well the LLM adapts to new tasks and challenges as the organization grows

- **Innovation and competitiveness**:

  - **Market position**: Analyzing improvements in market position due to the advanced capabilities provided by LLMs

  - **Product development**: The contribution of LLMs to the development of new products or enhancements to existing ones

ROI calculation

The ROI is typically calculated using the following formula:

ROI = (Financial Gains – Investment Cost)/Investment Cost x 100

Here, the following applies:

- **Financial gains**: This includes all the cost savings and additional revenue attributed to the use of LLMs

- **Investment cost**: This is the total cost of implementing and maintaining the LLMs

Conclusion

Assessing the ROI of LLMs is vital for organizations to determine the value brought by this technology. A positive ROI indicates that the LLMs are a beneficial investment, contributing to both financial health and operational excellence. However, ROI assessment is not just about the numbers—it also reflects strategic value, such as increased agility, customer satisfaction, and innovation, which may not be immediately quantifiable but are essential for long-term success.

Summary

In this chapter, we covered three model case studies. For our first model case study, Comet Communications implemented LLMs to improve customer service. The LLMs were integrated into existing systems for NLU and response generation, customized for the brand's voice, and deployed across multiple channels. Continuous learning and monitoring were essential, leading to improved customer satisfaction and operational efficiency. The LLMs were trained using historical customer service data, focusing on understanding company-specific terminology and frequent customer issues. A pilot program tested the LLMs, resulting in positive customer feedback, reduced response times, and lower operational costs, despite the need for continual updates and agent collaboration.

For our second model case study, Digimarket Corporation utilized LLMs for content creation and SEO, resulting in personalized marketing, increased engagement, and first-page search rankings. Challenges included maintaining content quality and brand consistency while avoiding algorithm bias.

TermCorp, our third model case study, applied LLMs to automate routine tasks, analyze financial documents, and optimize processes. This led to faster operations and improved data accuracy, as well as cost reductions. Integration with legacy systems and employee adaptation were initial hurdles.

Finally, the ROI of implementing LLMs was evaluated by analyzing cost savings, revenue growth, efficiency gains, and improved service quality, which demonstrated the financial and strategic benefits of LLM adoption.

In the next chapter, we will explore the ecosystem of LLM tools and frameworks.

13

The Ecosystem of LLM Tools and Frameworks

An exploration of the rich ecosystem of tools and frameworks available for **large language models (LLMs)** awaits you in this chapter. This exploration is crucial as it provides a detailed guide for selecting and integrating LLMs within existing tech stacks. We will offer a roadmap for navigating the selection of open source versus proprietary tools and comprehensively discuss how to integrate LLMs within existing tech stacks. The strategic role of cloud services in supporting NLP initiatives will also be unpacked.

In this chapter, we're going to cover the following main topics:

- Surveying the landscape of AI tools
- Open source versus proprietary – choosing the right tools
- Integrating LLMs with existing software stacks
- The role of cloud providers in NLP

By the end of this chapter, you should be equipped with a nuanced understanding of the AI tooling landscape and be capable of discerning open source and proprietary options for your specific needs. You'll have a clear guide on how to seamlessly integrate LLMs into your existing software stacks, and you'll grasp the pivotal role that cloud providers play in the realm of NLP.

Surveying the landscape of AI tools

LLMOps platforms streamline the deployment, fine-tuning, and management of LLMs, providing essential tools for enhancing their performance and integration across various applications. Here is an explanation of these AI tools:

- **LLMOps platforms**: These platforms are specifically designed for LLM operations or are extensions of existing MLOps platforms. They facilitate tasks such as fine-tuning and versioning for LLMs.

Here are some examples:

- **Cohere**: This is known for its user-friendly interface and LLM deployment solution
- **GooseAI**: This offers fine-tuning and deployment services for LLMs
- **Anthropic**: This is focused on generative AI, Anthropic aims to build safe and useful LLMs
- **OpenAI**: This offers pioneering research and development in LLMs

- **Integration frameworks**: These tools aid in developing LLM applications, such as document analyzers, code analyzers, and chatbots. They provide an interface for integrating LLMs into various applications.

  Notable frameworks include the following:

  - **LangChain**: This provides seamless integration for LLM-based applications
  - **Humanloop**: This enables efficient LLM integration with human feedback loops
  - **LlamaIndex**: LlamaIndex allows developers to query their private data using LLMs
  - **Orkes**: Orkes offers a workflow engine specifically designed for building complex LLM applications

- **Vector databases (VDs)**: VDs store high-dimensional data vectors, which can be useful for LLM operations.

  Here are some examples:

  - **Pinecone**: Pinecone offers a specialized VD system
  - **Weaviate**: Weaviate is another VD designed for semantic search and knowledge graph applications
  - **Qdrant**: Qdrant provides a high-performance VD for similarity search
  - **Milvus**: Milvus focuses on scalable vector storage and retrieval
  - **Vespa**: Vespa offers a versatile VD system
  - **Deep Lake**: A versatile VD for LLM-related tasks

- **Fine-tuning tools**: These frameworks or platforms allow the fine-tuning of pre-trained models. They streamline the process of modifying, retraining, and optimizing LLMs for specific tasks.

  Here are some examples:

  - **Hugging Face Transformers**: A popular library for fine-tuning and using pre-trained LLMs
  - **PyTorch**: This is widely used for LLM research and fine-tuning
  - **TensorFlow**: This offers LLM fine-tuning capabilities

- **Lakera**: Lakera provides a comprehensive guide to LLM fine-tuning, covering best practices, tools, and methods

- **Anyscale**: Anyscale showcases evolving tech stacks for LLM fine-tuning and serving

- **RLHF tools**: RLHF tools incorporate human feedback into the learning loop. They enhance LLM fine-tuning by incorporating large-scale data labeling and can be useful for AI governance.

  Here are some examples:

  - **Clickworker**: This leverages human input for LLM improvement

  - **Appen**: This provides data labeling and feedback for LLMs

  - **Scale AI**: This offers a data platform for AI with multiple annotation services, including image, sensor, and text data, to train and validate machine learning models

  - **Lionbridge** This specializes in data annotation and model training for AI

  - **Cogito**: This delivers a range of data annotation services, including sentiment analysis and intent recognition, for refining LLMs

Remember that the LLM landscape is dynamic, and new tools may emerge. These companies and tools collectively contribute to advancing language models across various domains.

Open source versus proprietary – choosing the right tools

When it comes to choosing the right tools for working with LLMs, one of the fundamental decisions is whether to use open source or proprietary software. Both choices come with their own sets of advantages and challenges that need to be considered based on the project requirements, budget, expertise, and long-term strategy.

Open source tools for LLMs

Using open source tools for LLMs has some advantages and disadvantages. We will explore them in detail in the following sections.

Advantages

Let's go through the advantages first.

Cost-effectiveness

Open source tools are inherently devoid of the licensing fees that accompany many proprietary software options. This characteristic is of paramount importance, particularly to entities operating under stringent budget constraints, such as start-ups, independent researchers, or educational institutions. The absence of a financial barrier to entry not only lowers the threshold for initial software deployment but also democratizes access to advanced computational tools such as LLMs.

Resource allocation without the burden of licensing fees allows for several strategic advantages:

- **Resource allocation**:

  - The savings accrued from the non-existence of purchase or subscription costs can be strategically redirected to bolster other facets of an operation.

  - In the area of hardware acquisition, the funds saved can be used to purchase better or more hardware, which is often a critical bottleneck in the performance of compute-intensive tasks such as those run by LLMs.

  - Human capital is arguably the most valuable asset in any technological venture. The funds conserved can be funneled into attracting and retaining talented individuals who can drive the project forward.

- **Encouragement of experimentation and innovation**:

  - Financial flexibility is a catalyst for innovation. When the barrier to entry is lowered, it opens the door for a broader range of experiments and projects that might not be feasible under financial constraints imposed by proprietary software costs.

  - Innovators and researchers can iterate rapidly, testing hypotheses and refining their models without the overhang of escalating costs. This agility can lead to faster discoveries and the rapid evolution of LLM capabilities.

Community support

In essence, community support in open source projects for LLMs is a powerful force that drives innovation, ensures the quality and security of the software, and fosters a collaborative environment where diverse ideas and solutions can flourish. It is an engine of collective intelligence that pushes the boundaries of what can be achieved in the field of AI.

Optimizing resource allocation provides several key benefits:

- **Collective pool of knowledge**:

  - Open source projects are often the nexus of collective intellectual effort. Developers and users from around the world contribute to a shared repository of knowledge, encompassing diverse perspectives and expertise.

- The community's wide-ranging expertise accelerates learning and skill development. Individuals can build upon a base of existing knowledge without starting from scratch, leading to more efficient progress in the field.

- **Faster problem resolution**:

  - The extensive network of support that characterizes open source projects can significantly expedite the problem-solving process. With many eyes on the same problem, there's a higher likelihood that someone has encountered and resolved similar issues.

  - Platforms such as forums, chat groups, and other online communities serve as real-time, dynamic support systems where individuals can seek assistance.

- **Enhanced robustness and security**:

  - The open source model invites scrutiny from anyone with an interest in the project, leading to more eyes reviewing the code. This process can lead to the identification and remediation of bugs and vulnerabilities that might otherwise go unnoticed in a closed-source environment.

  - A larger number of contributors can also mean a diversity of approaches to security, ensuring that the software is not just robust in its functionality but also in its defense against potential exploits.

- **Diversity of contributions**:

  - The open source ecosystem thrives on contributions that come from a variety of sources – individual hobbyists, academic researchers, corporate employees, and others. This diversity ensures that a broad range of use cases and viewpoints are considered during development.

  - Contributions can range from bug fixes and feature enhancements to security patches and performance improvements, all of which serve to strengthen the software.

- **Quality assurance (QA)**:

  - The iterative nature of open source development, compounded by community feedback, tends to yield high-quality software. Users and developers alike are constantly testing, fixing, and updating the code, which often leads to software that is both refined and resilient.

  - The software evolves not just through planned updates but through continuous, incremental improvements and audits by the community.

- **Sustainability and longevity**:

  - Community support can contribute to the sustainability of open source projects. A vibrant, active community can continue development even if the original creators are no longer involved, ensuring the longevity of the project.

- The project's sustainability is also underpinned by the fact that it does not rely on the financial success or strategic direction of a single company, but on the collective will of its contributors.

Transparency

The **transparency** inherent in open source LLMs is a multifaceted advantage that spans trust, security, compliance, and ethics. It provides a comprehensive suite of benefits that can lead to more responsible and reliable AI systems, thereby fostering a greater level of trust among users, developers, and the broader society in which these technologies are deployed.

The open source nature of software provides several transformative benefits:

- **Full transparency in the code base**:

  - Open source software is synonymous with an open-book policy regarding its code base. This level of transparency allows any user, developer, or researcher complete visibility into the internal workings of the software.

  - For LLMs, which are complex and often operate as black boxes, having an open code base demystifies the process by which they operate. It can empower users to tweak and improve the models, ensuring that the outcomes are explainable and aligned with expectations.

- **The foundation of trust and security**:

  - Transparency is a cornerstone of trust in software systems. When LLMs are used in critical applications, such as healthcare diagnostics, financial forecasting, or personal data processing, the stakes are incredibly high. In these scenarios, it's imperative that the models behave in predictable and secure ways.

  - Open source transparency assures users that there are no hidden processes that could potentially mislead or harm the end user. It also means that any security measures are open for inspection and critique, allowing for a more robust security posture.

- **Facilitation of audits and verifications**:

  - In many industries, software systems are subject to strict regulatory compliance standards. The open nature of the source code in open source LLMs enables comprehensive audits by third parties, who can verify that the software adheres to industry regulations and standards.

  - This is especially pertinent in fields such as healthcare or finance, where software systems need to comply with regulatory frameworks such as the **Health Insurance Portability and Accountability Act** (**HIPAA**) or Sarbanes-Oxley. Auditors can inspect the code to ensure that the software meets all necessary compliance requirements.

- **Enhanced credibility with stakeholders**:

    - Transparency not only builds trust with users but also enhances the credibility of the software among stakeholders. When investors, partners, or regulators can see that an organization uses transparent and verifiable LLMs, it can facilitate smoother partnerships, funding opportunities, and regulatory approvals.

- **Community-driven security enhancements**:

    - The open source model invites a communal approach to security. Since the source code is available to everyone, it benefits from the collective vigilance of a broad community of experts who can spot and rectify security flaws.

- **Support for ethical AI development**:

    - As the field of AI continues to grapple with ethical concerns, transparency in LLMs provides a framework for ethical oversight.

    - This level of openness is crucial for the responsible development of AI systems, ensuring that they are fair, unbiased, and aligned with societal values.

Flexibility and customization

The flexibility and customization potential of open source software, particularly for LLMs, provide a robust foundation for innovation and adaptation. Organizations can craft a software solution that not only meets their current needs but can also evolve alongside their ambitions and challenges, all while fostering a dynamic and collaborative development environment.

Open source software offers significant adaptability and scalability, providing critical advantages in several ways:

- **Tailoring to specific needs**:

    - Access to the source code is like having a master key to the software; it allows developers to get into the very heart of the program and make adjustments that fit their unique requirements. This is an immense advantage when the application of LLMs extends into niche domains with specialized needs that off-the-shelf products cannot meet.

    - Customization can range from simple user interface tweaks to complex alterations in the algorithms and processing pipelines.

- **Scalability**:

    - As project requirements evolve, the need to scale the software can arise. Open source software can be modified to handle an increase in workload, such as larger datasets or more complex queries, without the need for a complete overhaul of the system.

- Scalability can also refer to the ability to enhance the software's performance efficiency, enabling faster processing times and more economical use of computational resources, which is critical for the data-intensive tasks that LLMs perform.

Rapid development and innovation

The open source model provides a fertile ground for rapid development and innovation in the field of LLMs. By leveraging the collective efforts of a global community, the development of LLMs can proceed at an unprecedented pace, with a multitude of contributors driving the technology forward in a variety of creative and unexpected ways.

Open source projects benefit immensely from collaborative efforts, offering several key advantages:

- **Accelerated evolution through collaborative contributions**:

  - Open source projects are unique in that they harness the collective capabilities of a diverse and global developer community. Each contributor can bring their own insights, skills, and experiences to the project, which can lead to a compounding effect on the speed of development and the introduction of innovative features.

  - For LLMs, the rapid inclusion of improvements—from language support enhancements to algorithmic efficiency gains—can be integrated into the project as soon as they are developed.

- **Collaborative approach leading to novel solutions**:

  - Open source development is inherently collaborative, not competitive. This environment fosters a culture where sharing knowledge is the norm, and it's this culture that leads to the discovery of novel approaches and techniques.

  - With LLMs, such collaboration might manifest as shared datasets, innovative training methods, or new neural network architectures. When these resources are shared, they can be tested, refined, and potentially integrated by others into a variety of projects, thereby enriching the entire ecosystem.

- **Diversity of thought and experimentation**:

  - The diversity of the open source community is one of its greatest strengths. Individuals from different backgrounds and with different objectives contribute to the project, bringing a wide range of ideas to the table. This diversity encourages experimentation and can lead to breakthroughs that might not occur in more homogeneous groups.

No vendor lock-in

Avoiding **vendor lock-in** by using open source tools provides a strategic advantage, offering cost savings, technological agility, and the freedom to innovate. It empowers organizations to make decisions based on technical merit and strategic fit rather than being constrained by the decisions of a single vendor. This is especially critical in the rapidly evolving field of LLMs, where flexibility and the ability to quickly adapt to new developments can provide a significant competitive edge.

Avoiding vendor lock-in offers numerous benefits:

- **Avoidance of vendor lock-in:**

  - Substantial switching costs lead to vendor lock-in, where a customer becomes reliant on a specific vendor for products and services and finds it difficult to transition to another vendor

- **Cost implications:**

  - Vendor lock-in is often associated with escalating costs over time. As a vendor's product becomes more embedded in an organization's infrastructure, the vendor gains leverage to increase prices or change terms of service in ways that can be unfavorable to the customer.

  - Open source software, by contrast, is usually free from such constraints, which can result in significant long-term cost savings and budgetary predictability.

- **Agility and flexibility in technology choices:**

  - The tech industry is characterized by rapid evolution, and the flexibility to adapt to new technologies is crucial. Being locked into a single vendor's ecosystem can hinder an organization's ability to adopt new, more efficient, or cost-effective technologies.

- **Reduced risk of incompatibility and transition costs:**

  - Proprietary solutions often use closed formats and protocols, which can lead to compatibility issues when transitioning to another system. Open source tools, however, tend to favor open standards, which can minimize these risks. Additionally, if a vendor goes out of business or discontinues a product, customers can be left with unsupported software and a potentially expensive migration process.

Wider adoption and collaboration

The lack of financial barriers in open source software leads to wider adoption, which in turn fosters a rich environment for collaboration and innovation. This collaborative ecosystem is conducive to the rigorous testing and continual refinement of tools, encouraging breakthroughs and ensuring the sustainability and evolution of LLMs in a way that proprietary models may not.

The elimination of financial barriers in open source software facilitates a range of benefits:

- **Facilitation of wider adoption**: Open source software, with its typically non-existent price tag, removes a significant barrier to entry. Without the financial burden, a broader demographic, from solo developers to large enterprises, can access the technology. This inclusivity not only increases the user base but also brings a variety of perspectives and skills to bear on the software's usage and development.

- **Enhanced testing and refinement**: A large and diverse user base can contribute to the rigorous testing of open source tools. In the context of LLMs, where different languages, dialects, and textual formats can vastly affect performance, such extensive testing is invaluable.

- **Promotion of collaborative innovation**: When financial barriers are removed, it encourages not just adoption but active collaboration. Academics, industry professionals, and hobbyists are able to contribute to the project, bringing together a rich tapestry of knowledge and expertise.

The adoption of open source tools for LLMs can present several strategic advantages, from cost savings to innovation and collaboration, making them a valuable resource for anyone in the field of AI and machine learning.

Challenges

Apart from these advantages, there are also some limitations associated. Let's have a look.

Resource intensity

While open source tools for LLMs offer many benefits, they also present challenges in terms of resource intensity. Implementing and maintaining these tools demands time, expertise, and often an investment in infrastructure and training. These indirect costs need to be carefully considered and managed to fully leverage the advantages of open source LLMs without encountering prohibitive barriers to their effective use:

- **Expertise and time investment**:

  - Open source software can often be complex and may not come with the same level of out-of-the-box readiness or comprehensive documentation that one expects from commercial software. Implementing these solutions effectively may require a high level of technical expertise and a willingness to invest significant time.

  - With LLMs, this challenge is pronounced due to the sophistication of the technology and the specialized knowledge required to not only implement but also to train, fine-tune, and maintain these models. Individuals or organizations may need to invest in training or hiring skilled personnel, which can be a significant indirect cost.

- **Maintenance demands**:

  - Unlike proprietary software that often includes vendor support and regular updates, maintenance of open source tools typically falls to the users. This includes updating the software, patching vulnerabilities, and ensuring compatibility with other systems and dependencies.

  - For LLMs, maintenance is particularly resource-intensive because the field is rapidly advancing, meaning that keeping up with the latest developments and integrating them can be a continuous and demanding task.

- **Hidden costs**:

  - While the software itself may be free, there are often hidden costs associated with open source tools. These can include the need for additional software or hardware to support the tool, training costs for staff, and the potential need for paid support or consultancy to fill gaps in expertise.

  - With LLMs, these hidden costs can accumulate rapidly, especially considering the data processing and computational power required to run these models effectively.

Support and reliability

While the collaborative nature of community support is a hallmark of open source software, the absence of dedicated, professional support can pose challenges in terms of reliability and the timely resolution of issues. This is particularly relevant for organizations using LLMs for critical applications, where the cost of failure can be high.

While open source software offers many advantages, it also presents unique challenges:

- **Variability in community support**:

  - The support system for open source software is generally community-driven, which means that the quality and speed of support can vary greatly. While there are often active forums and user groups, there is no guarantee of timely assistance, and the level of expertise may be inconsistent.

  - In the context of LLMs, which are complex systems requiring sophisticated understanding, the absence of guaranteed professional support can be a significant risk. If an organization encounters a specialized issue or requires immediate help, community forums may not provide the level of service needed.

- **Professional support services**:

  - Proprietary solutions often come with the option of a **service-level agreement** (**SLA**), ensuring a certain standard of support. Open source tools typically do not offer this level of dedicated support as part of the package, which can lead to challenges, especially in a production environment where downtime or unresolved issues can have serious implications.

- Organizations using open source LLMs may need to rely on third-party vendors for professional support, which can introduce additional costs and complexities. Alternatively, they may need to build their own internal expertise, which can be a costly and time-consuming process.

- **Reliability and accountability**:

  - With proprietary software, there is a clear line of accountability to the vendor for the performance and reliability of the product. In the open source world, the software is often the result of contributions from many different individuals and organizations, which can make accountability diffuse.

  - For critical applications of LLMs, the lack of a single point of accountability can be a significant concern. If a system fails or does not perform as required, it may be challenging to identify a responsible party to address the issue.

- **Ongoing development and updates**:

  - The continuity of development for open source software can be uncertain. While some projects are robustly maintained, others may suffer from periods of stagnation or may be abandoned altogether if the community's interest wanes or key contributors leave.

  - For LLMs, where ongoing development is crucial to keep pace with the latest advancements in the field, the lack of reliable updates can limit the software's usefulness over time.

- **QA processes**:

  - Open source projects may not have the same rigorous QA processes as commercial software. While the community can and often does contribute to testing and QA, the processes can be less structured and comprehensive than those provided by a dedicated vendor team.

  - This can impact the reliability of LLMs, where the accuracy and quality of the model's outputs are paramount.

- **Custom solutions and workarounds**:

  - In the absence of dedicated support, organizations may find themselves having to develop custom solutions or workarounds to problems. This can be a drain on resources and may not always lead to the most efficient or effective outcomes.

  - With LLMs, which may be integrated into larger systems, developing these solutions can be particularly complex and require a deep understanding of both the model and the system architecture.

Integration

While open source tools offer many advantages, they can present challenges when it comes to integration with existing systems. Compatibility issues, the need for custom development, and the potential lack of enterprise-ready features are all factors that organizations must consider. Successful integration of LLMs into existing IT infrastructures requires careful planning, a clear understanding of both the open source tool and the target environment, and a potentially significant investment in development and testing.

Integrating open source tools, especially complex systems such as LLMs, can present several challenges:

- **Compatibility issues**:

  - Open source tools are developed by diverse groups of contributors and may not always adhere to standardized protocols or interfaces, which can lead to compatibility issues with existing systems. This is especially relevant for LLMs, as they often need to interact with various data sources, processing pipelines, and application interfaces.

  - Ensuring that an open source LLM works harmoniously with proprietary or legacy systems can require significant effort in terms of developing middleware or custom adapters. Such compatibility layers may need to be built from scratch, requiring in-depth knowledge of both the open source software and the existing systems' architecture.

- **Seamless integration challenges**:

  - Proprietary tools are often designed with integration in mind, offering built-in connectors and plugins for popular enterprise software. Open source tools, in contrast, may lack these turnkey integration solutions, potentially leading to more complex and labor-intensive integration processes.

  - For LLMs, which are data-driven and may need to integrate tightly with content management systems, databases, or other AI services, the lack of seamless integration can be a significant hurdle. Organizations may need to allocate more resources to ensure the smooth flow of data and functionality between systems.

- **Documentation and support for integration**:

  - The quality and comprehensiveness of documentation can vary widely in the open source community. While some projects may provide extensive integration guides, others may have sparse or outdated documentation, which can complicate integration efforts.

  - Without adequate documentation, developers may need to rely on trial and error or seek guidance from community forums, which can be time-consuming and may not yield definitive solutions for integrating LLMs with specific systems or technologies.

- **Evolving landscapes and standards**:

  - The IT landscape is continuously evolving, with new standards and best practices emerging regularly. Open source tools may lag in adopting these new standards, or they may adopt them in ways that are not consistent with industry norms, further complicating integration efforts.

  - For LLMs, staying abreast of data privacy standards, security protocols, and API conventions is critical. Any lag in alignment can be problematic when trying to integrate with systems that adhere to the latest standards.

Proprietary tools for LLMs

After reviewing the advantages and challenges of using open source tools for LLMs, it's time to turn to proprietary tools.

Advantages

Let's start with their advantages.

Ease of use

The advantages of proprietary tools for LLMs in terms of ease of use stem from a user-centric design philosophy, comprehensive support infrastructure, and a focus on providing a reliable and professional-grade product. These factors contribute to a smoother user experience, making proprietary tools attractive to individuals and organizations looking for turnkey solutions that allow them to leverage the power of LLMs with minimal setup and ongoing maintenance efforts.

Proprietary tools, particularly in the realm of LLMs, offer a suite of user-centric features that enhance accessibility and ease of use:

- **User-friendly interface**:

  - Proprietary tools are typically developed with a focus on user experience, offering interfaces that are intuitive and visually appealing. These interfaces are often the result of substantial research and user testing to ensure that they meet the needs of a broad user base.

  - For LLMs, this means providing access to complex functionalities through simplified dashboards, clear menu structures, and comprehensive onboarding processes. It enables users with varying levels of technical expertise to work with the model without needing to understand the underlying code.

- **Out-of-the-box functionality**:

  - One of the key selling points of proprietary software is its readiness for immediate use upon installation, with minimal setup required. This contrasts with open source tools that may require additional configuration or the installation of dependencies before they can be used effectively.

  - Proprietary LLMs are likely to come pre-configured with a range of defaults that are suitable for many common applications, allowing users to start their tasks with minimal delay.

- **Streamlined workflows**:

  - Proprietary LLMs often include streamlined workflows that guide users through the process of using the tool, from data input to model training and output analysis. This can significantly reduce the learning curve and increase productivity.

  - These workflows are also often accompanied by wizards or help functions that can guide users through complex processes step by step, making the technology accessible to non-experts.

- **Comprehensive documentation and training**:

  - Vendors of proprietary software usually provide extensive documentation, tutorials, and training materials. These resources are designed to help users get the most out of the software and can be crucial in helping to overcome any initial barriers to effective use.

  - For LLMs, which can be complex to operate, having well-structured and easily accessible documentation can be a significant advantage, enabling users to understand and leverage the tool's full capabilities.

- **Support services**:

  - Proprietary software often comes with customer support services included in the purchase price or available as an added service. This professional support can range from troubleshooting assistance to help with customization or integration.

  - Users of proprietary LLMs can typically rely on a consistent level of support, ensuring that any issues can be resolved quickly, which is essential for maintaining continuity in business operations.

- **QA and reliability**:

  - Proprietary software vendors have reputations to uphold and are therefore motivated to ensure their products meet high standards of quality and reliability. Extensive testing before release is standard practice.

  - Users of proprietary LLMs can expect a product that has been vetted for a wide range of scenarios and is less likely to contain critical bugs or errors.

Support

The professional support and regular updates that come with proprietary tools for LLMs are significant advantages. They ensure that users have access to expert assistance, continuous improvements, and reliable maintenance of their software, which can be particularly valuable for organizations that rely on the functionality and performance of LLMs for their core operations.

Proprietary software vendors provide comprehensive professional support services that offer several key benefits for users:

- **Professional support services**:

  - Proprietary software vendors typically offer a range of support services, which can be essential for users who rely on LLMs for important business functions. This support can come in various forms, including direct access to technical experts, help desks, and customer service centers.

  - The professional support teams are usually well trained on the specific software and can provide quick, reliable assistance, which can be crucial when time-sensitive issues arise. This level of support is particularly valuable for organizations that may not have in-house expertise in the complexities of LLM technology.

- **SLAs**:

  - Proprietary vendors often work with SLAs that guarantee a certain level of service, response time, and availability. This contractual assurance can be critical for businesses that depend on LLMs for critical operations.

  - SLAs provide peace of mind and a level of predictability for businesses, ensuring that they know what to expect in terms of support and service quality.

Stability and reliability

The stability and reliability of proprietary tools for LLMs stem from structured development and release processes, stringent QA, and a focus on ensuring that updates improve the software without causing unnecessary disruption. This creates an environment where businesses can rely on their LLMs to perform consistently and effectively over time, providing peace of mind and allowing for long-term planning and investment in these tools.

Proprietary software vendors maintain stable release cycles and rigorous QA processes, providing numerous advantages for users of proprietary LLMs:

- **Stable release cycles**:

  - Proprietary software vendors often have well-established release cycles for updates and new versions of their software. This controlled release process is designed to ensure that each update is thoroughly tested and stable before it's made available to customers.

- For users of proprietary LLMs, this means they can expect a platform that remains consistent over time, with updates that are less likely to introduce significant changes that would require users to alter their workflows or retrain their models extensively.

- **QA processes**:

  - Before releasing any updates, proprietary tools undergo rigorous QA testing. The QA process in a proprietary setting is systematic and comprehensive, aiming to catch and fix any potential issues before the software reaches the customer.

  - This focus on quality leads to a more reliable and stable experience for users of proprietary LLMs, reducing the likelihood of encountering bugs or other issues that could disrupt their operations.

- **Predictable performance**:

  - Proprietary LLMs are engineered to provide predictable performance across various conditions and use cases. The providers ensure that the models perform optimally within the expected parameters, providing a level of reliability that is essential for users who rely on these tools for critical decision-making.

  - The reliability of proprietary LLMs is especially crucial in environments where the cost of failure is high, such as finance, healthcare, or legal industries.

- **Long-term support (LTS) versions**:

  - Many proprietary vendors offer LTS versions of their software, which receive maintenance updates for an extended period. These LTS versions are ideal for enterprise environments where stability is more important than having the latest features.

  - Users leveraging proprietary LLMs for core business functions can benefit from LTS versions, which provide the security of continued support without the need for frequent upgrades.

Compliance and security

The commitment to compliance and security is a key advantage of proprietary tools for LLMs. Vendors who invest in these areas help ensure that their tools not only protect sensitive data but also meet the regulatory requirements that are critical for sensitive applications. This support can provide peace of mind for organizations and reduce the burden of managing compliance and security risks internally.

Proprietary vendors ensure their software adheres to industry standards and regulations, offering several crucial benefits:

- **Adherence to industry standards**:

  - Proprietary vendors typically design their tools to comply with industry standards and regulations. This includes following best practices for data handling, privacy, and security measures, which are crucial for maintaining the integrity and confidentiality of sensitive information.

  - For LLMs, this means that the software is more likely to be aligned with standards such as GDPR for data protection, HIPAA for healthcare information, and PCI Data Security Standard for payment data security, among others.

- **Certifications and audits**:

  - Proprietary software vendors often undergo regular third-party audits and strive to obtain certifications that affirm their compliance with various industry standards. These certifications serve as evidence of the software's reliability and adherence to regulatory requirements.

  - For organizations using LLMs, these certifications can simplify compliance efforts, as they can rely on the vendor's software to meet the necessary legal and industry-specific regulatory frameworks.

- **Security features**:

  - Security is a paramount concern in proprietary software development. Vendors invest in building robust security features, such as encryption, access controls, and activity monitoring, to protect against unauthorized access and data breaches.

  - In the context of LLMs, which process and generate large volumes of data, including potentially personal or proprietary information, these security features are essential to protect the data and the insights derived from it.

- **Risk mitigation**:

  - By providing tools that are compliant with regulations and standards, proprietary vendors help mitigate the risks associated with non-compliance, such as legal penalties, data breaches, and reputational damage.

  - Users of proprietary LLMs can leverage the vendor's expertise in compliance to reduce their own risk exposure, especially when operating in sectors where the mishandling of data can have serious consequences.

Challenges

Let's also have an overview of the associated challenges.

Cost

While proprietary tools for LLMs offer numerous benefits in terms of support, reliability, and compliance, they can also represent a substantial financial investment. Organizations must carefully evaluate the direct and indirect costs of these tools, including licensing and subscription fees, additional services, and potential costs related to scaling and customization, to determine whether they align with the organization's budget and long-term financial planning.

Proprietary software, including LLMs, often involves various costs:

- **Licensing fees**:

  - Proprietary software typically requires the purchase of a license to use the software. These licenses can be structured in various ways, such as per user, per machine, or per core/CPU, and may vary greatly in cost depending on the scale of the deployment.

  - For LLMs, the licensing fees might also be calculated based on the volume of data processed or the number of API calls made, which can add to the overall cost, especially for organizations handling large quantities of data.

- **Subscription fees**:

  - Many proprietary LLMs are offered under a subscription model, where users pay a recurring fee to use the software. Subscriptions can provide access to a suite of services and ensure that the software stays up to date with the latest features and security updates.

  - While subscription models can lower the initial cost of entry compared to perpetual licenses, over time, they can amount to a significant expenditure, especially if the subscription includes tiered pricing based on usage levels.

- **Additional services or add-ons**:

  - Proprietary vendors often offer a range of additional services and add-ons that can enhance the functionality of the software. These might include advanced analytics, custom model training, premium customer support, and more.

  - These services can be essential for making the most of LLMs but come at an additional cost. Organizations may find that the base version of the software requires several add-ons to meet their specific needs, which can significantly increase the total cost of ownership.

- **Integration and customization costs**:

  - While proprietary tools may offer ease of use and stability, integrating them with other systems or customizing them to suit specific requirements can require additional investment in professional services or custom development work.

  - For LLMs, integration with existing databases, CRM systems, or other enterprise software may necessitate specialized services that add to the overall cost.

Less flexibility

While proprietary tools for LLMs provide benefits such as ease of use, support, and stability, they often lack the flexibility that some users require. This can manifest in constraints on customization, integration challenges, and a limited ability to adapt the tool to specific needs without incurring additional costs or relying on the vendor's development schedule. Organizations must weigh these factors against their need for a tailored solution when considering proprietary LLMs.

Proprietary software, including LLMs, presents certain limitations and dependencies:

- **Customization limitations**:

  - Proprietary software is often designed as a closed system, with limited options for customization. This is because the source code is not accessible for users to modify, which is a stark contrast to open source software where customization is a key feature.

  - In the case of LLMs, this means that users might not be able to adjust or extend the model's architecture, tweak its learning algorithms, modify its interface to suit their unique workflows, or integrate with their existing systems seamlessly.

- **Dependence on the vendor for enhancements**:

  - When customization is needed for proprietary LLMs, users are typically dependent on the vendor to provide these enhancements. This can result in waiting for the vendor to develop a requested feature or change, which may not always align with the user's timeline or priorities.

  - Additionally, vendors may prioritize developments based on their strategic interests or the interests of their largest customers, potentially leaving smaller users with unmet customization needs.

- **Integration with other systems**:

  - Proprietary LLMs may not integrate smoothly with other systems, particularly if those systems are from different vendors or are built on open source platforms. This can force organizations to work within a more rigid framework, using only the tools and integrations that the vendor supports.

- Overcoming these integration challenges often requires the use of APIs, middleware, or other interfacing tools that the vendor provides, which may not offer the level of control or data interaction that the user desires.

Choosing between open source and proprietary tools for LLMS

Choosing between open source and proprietary tools for LLMs will depend on several factors:

- **Project budget and resources**: If budgets are tight and in-house expertise is available, open source may be the way to go. For organizations that prefer a more managed solution and can afford it, proprietary may be more suitable.

- **Customization needs**: If the project requires heavy customization, open source tools may offer the necessary flexibility.

- **Scalability and integration**: For projects that need to scale quickly and integrate with other systems, proprietary tools might offer more robust solutions.

- **Security and compliance**: For projects that handle sensitive data or require strict compliance with regulations, proprietary solutions often provide comprehensive security features and compliance certifications.

Ultimately, the decision may not be binary, and many organizations find that a hybrid approach—using a mix of open source and proprietary tools—best meets their needs. It's also common to start with open source tools for prototyping and experimentation, and then switch to proprietary solutions for production-level deployment.

In conclusion, the choice between open source and proprietary tools for LLMs should be guided by a clear understanding of the project requirements, available resources, and strategic goals. It's also important to stay informed about the evolving landscape of LLM tools and to periodically reassess the tooling strategy as new technologies and updates become available.

Integrating LLMs with existing software stacks

Integrating LLMs with existing software stacks is an important step for businesses and developers looking to leverage the power of advanced NLP within their current technological ecosystem. This integration process typically involves several key considerations:

- **Assessment of requirements**: Understanding the specific needs of the business or application is crucial. This includes determining what tasks the LLM will perform, such as text generation, sentiment analysis, or language translation.

- **Choosing the right LLM**: Depending on the requirements, a suitable LLM should be chosen. For example, GPT-4 might be chosen for its text generation capabilities, while BERT might be preferred for its performance in understanding context in search queries.

- **APIs and integration points**: Most LLMs provide APIs that are the primary means of integration with existing software stacks. These APIs allow the LLM to communicate with other systems, passing data back and forth as needed.

- **Data handling and processing**: To effectively integrate an LLM, you need to ensure that your data is in the right format. This might involve preprocessing steps to clean and structure the data before it can be used by the LLM.

- **Infrastructure considerations**: LLMs can be resource-intensive, so it's important to ensure that your existing infrastructure can handle the additional load. This may involve scaling up your servers or moving to a cloud-based solution.

- **Security and privacy**: When integrating an LLM into your software stack, you need to consider the security and privacy implications, particularly if you're dealing with sensitive or personal data. This might involve implementing additional security measures or ensuring that the data is anonymized before being processed by the LLM.

- **Compliance and ethics**: It's vital to ensure that the use of LLMs complies with relevant laws and regulations, such as GDPR for data protection. Ethical considerations should also be taken into account, ensuring that the LLM is used in a manner that is fair and does not perpetuate biases.

- **Testing and validation**: Before fully integrating an LLM into your software stack, it should be thoroughly tested. This testing should validate that the LLM performs as expected and works seamlessly with the other components of your software stack.

- **Monitoring and maintenance**: Once integrated, the LLM should be continuously monitored to ensure it is functioning correctly. Regular maintenance may also be required to update the model or the integration as new versions are released.

- **User training**: It's often necessary to train users on how to interact with the LLM, especially if they are using it as part of their workflow, such as customer service representatives or content creators.

- **Scalability and futureproofing**: The integration should be designed in a way that it can scale as the usage of the LLM grows. Also, it should be flexible enough to accommodate future advancements in LLMs.

- **Documentation**: Comprehensive documentation of the integration process and the way the LLM interacts with other system components is important for maintenance and future reference.

Integration of LLMs into existing software stacks is not a one-size-fits-all process. It requires careful planning, consideration of technical and ethical implications, and ongoing management. However, when done correctly, it can significantly enhance the capabilities of your software stack and provide valuable services to your users.

The role of cloud providers in NLP

Cloud providers play a crucial role in the field of NLP by offering a wide range of services that democratize access to cutting-edge technologies. Their contribution can be categorized into several key areas:

- **Infrastructure**: Cloud providers offer the necessary computational infrastructure required for NLP tasks, which often demand significant processing power. This infrastructure supports the training of large-scale language models and the deployment of NLP applications without the need for organizations to invest in their own hardware.

- **Platform as a service (PaaS)**: Through PaaS offerings, cloud providers supply platforms that allow developers to build, deploy, and manage NLP applications without the complexity of building and maintaining the infrastructure.

- **NLP services and APIs**: Cloud providers such as Amazon AWS, Google Cloud Platform, and Microsoft Azure offer a suite of pre-built NLP services and APIs. These include text analysis, translation, sentiment analysis, and chatbot services, making it easier for businesses to integrate NLP capabilities into their applications.

- **ML frameworks and tools**: They provide access to ML frameworks such as TensorFlow, PyTorch, and MXNet, which are essential for building custom NLP models. These tools come with the added benefit of cloud scalability and managed services.

- **Data storage and management**: NLP models require access to vast datasets. Cloud providers offer scalable and secure data storage solutions, along with tools for managing and processing this data effectively.

- **AutoML and custom model training**: For organizations without the expertise to build models from scratch, cloud providers offer AutoML services, which automate the creation of custom NLP models tailored to specific needs.

- **Marketplaces for AI models**: Cloud platforms often have marketplaces where users can find and deploy pre-built NLP models, which can be further customized for specific tasks.

- **Security and compliance**: Cloud providers ensure that the deployment of NLP applications complies with security standards and privacy regulations, which is particularly important when processing sensitive data.

- **Global reach and localization**: They facilitate the deployment of NLP applications across global infrastructures, ensuring low latency and compliance with local data residency requirements. This is especially important for NLP applications that need to be localized for different languages and regions.

- **Research and development**: Cloud providers invest heavily in research, developing state-of-the-art NLP technologies that can be leveraged by their customers. They also often provide credits and support for academic research in NLP.

- **Community and support**: They foster a community of developers and provide extensive documentation, tutorials, and forums for support, which is invaluable for teams working on NLP projects.

- **Scalability and flexibility**: One of the most significant advantages of cloud providers in NLP is the ability to scale resources up or down as needed, providing flexibility and cost control for businesses of all sizes.

In conclusion, cloud providers are integral to the NLP ecosystem, providing the tools, services, and infrastructure that enable businesses to harness the power of language processing. They continually push the boundaries of what's possible in NLP, offering increasingly sophisticated solutions that allow for innovation and expansion in the field.

Summary

This chapter provided a comprehensive guide on the ecosystem of LLM tools, presenting critical insights for choosing between open source and proprietary options based on budget, customizability, and the need for support. It outlined the practicalities of integrating LLMs into existing software ecosystems and underscored the essential role of cloud providers in offering infrastructure, platforms, and services for NLP.

LLMOps platforms such as Cohere and OpenAI are vital for fine-tuning and deploying LLMs, whereas tools such as Hugging Face Transformers are crucial for model fine-tuning. RLHF tools, offered by entities such as Appen, enhance model training with human feedback.

The decision to adopt open source or proprietary tools must be informed by the specific needs, strategic goals, and resource availability of the organization. Cloud providers were highlighted as critical enablers, providing the necessary computational power and services to support NLP applications.

Putting it together, this chapter served as a decision-making roadmap for integrating LLMs and prepared you to navigate the ongoing developments in the field, anticipating future advancements such as GPT-5.

In the next chapter, we will review how you should prepare for GPT-5 and beyond.

14

Preparing for GPT-5 and Beyond

Anticipating future developments in the LLM space, we will prepare you for the arrival of GPT-5 and subsequent models in this chapter. We will cover the expected features, infrastructure needs, and skillset preparations. We will also challenge you to think strategically about potential breakthroughs and how to stay ahead of the curve in a rapidly advancing field.

In this chapter, we're going to cover the following main topics:

- What to expect from the next generation of LLMs
- Getting ready for GPT-5 – infrastructure and skillsets
- Potential breakthroughs and challenges ahead
- Strategic planning for future LLMs

By the end of this chapter, you should be equipped with the knowledge and foresight to anticipate future developments in the LLM space, particularly with the imminent arrival of GPT-5 and subsequent models.

What to expect from the next generation of LLMs

Looking ahead, LLMs will significantly improve in understanding and contextualizing information. Future models will maintain context over long interactions, ensuring coherent and fluid dialogue by integrating dynamic memory to recall past exchanges. This enhanced contextual capability will help LLMs resolve ambiguities based on broader conversations, making them adept at generating relevant and coherent content. These advancements will enrich user experiences and broaden the application of LLMs in complex, language-intensive scenarios, making them more proficient and helpful conversational partners across various domains. Let's explore these features in detail.

Enhanced understanding and contextualization

As we look to the future of LLMs, one of the most exciting developments is the enhanced understanding and contextualization capabilities that these models are expected to have. Here are several areas in which these improvements may manifest:

- **Long-term context management**: Future LLMs may be able to maintain context over longer interactions, remembering and referencing past parts of a conversation. This would allow for more natural and fluid dialogue, as the model wouldn't "forget" previous exchanges.

- **Dynamic memory integration**: By incorporating a dynamic memory component, LLMs could store and retrieve information from earlier in the conversation or from past interactions with the same user. This is a step beyond static contextual understanding, where the model can only reference immediate or recent input.

- **Cross-session learning**: Beyond a single session, LLMs might be capable of cross-session learning, where they remember user preferences, interests, and history across multiple interactions. This would lead to highly personalized experiences.

- **Contextual disambiguation**: Improved contextualization would enable LLMs to better understand ambiguous language based on a conversation's context. They could resolve ambiguities by considering the broader topic or by recalling how similar phrases were used previously in the dialogue.

- **Coherence and cohesion**: Coherence (the logical connection between ideas) and cohesion (how sentences and parts of text connect) in a model's responses are expected to improve, making conversations with LLMs more logical and easier to follow.

- **Advanced reference capabilities**: The ability to reference and understand indirect language, such as pronouns or elliptical constructions, will likely be enhanced, allowing for more complex and varied dialogue that remains clear and connected.

- **Context-aware content generation**: In content generation tasks, LLMs would be capable of creating text that is not only relevant to the immediate prompt but also considers the wider context, such as a user's known interests or the current cultural or social milieu.

- **Emotion and sentiment understanding**: By better understanding the context, LLMs could more accurately interpret the emotional tone and sentiment of user input, leading to more empathetic and appropriate responses.

- **Situational awareness**: Future models might develop situational awareness, allowing them to adapt their responses based on the perceived situation or setting of an interaction, such as a formal business meeting versus a casual chat.

- **Personalized learning paths**: In educational applications, LLMs could create personalized learning paths that consider a student's progress, interests, and the context of past performance.

- **Ethical and cultural sensitivity**: With better contextualization, LLMs could show improved sensitivity to ethical considerations and cultural contexts, avoiding misunderstandings and inappropriate responses.

Enhancing these capabilities will improve user experience and expand LLM applications in complex, language-intensive scenarios. As technology advances, LLMs will become more adept at sustaining meaningful and helpful interactions across various domains.

Improved language and multimodal abilities

The next generation of LLMs is expected to exhibit not only advanced language processing capabilities but also multimodal abilities. Here's a detailed look at what this might involve:

- **Multimodal processing**: Multimodal LLMs will be able to process and understand information from various data types beyond text, including visual elements (images and videos), auditory signals (speech and sounds), and possibly tactile feedback, allowing for richer interactions.

- **Content generation across media**: These models could generate coherent content that spans multiple forms of media. For example, an LLM might create a textual description that aligns with visual content or vice versa.

- **Contextual understanding of visual elements**: Improved language models will likely have a deeper understanding of the context within images and videos. This means recognizing objects, actions, and sentiments in visual media and relating them to textual information.

- **Advanced natural language generation (NLG)**: Coupled with visual or auditory data, NLG in multimodal LLMs will produce more descriptive and accurate narratives or captions, enhancing the storytelling aspect of content creation.

- **Audio and speech processing**: In audio processing, future LLMs could transcribe, interpret, and even generate human-like speech. This could revolutionize virtual assistants, making them more natural and responsive to vocal nuances.

- **Video understanding and generation**: LLMs might be capable of analyzing video content, understanding a sequence of events, and generating summaries or interactive elements based on video data.

- **Cross-modal translation**: One of the more intriguing prospects is the translation of content from one modality to another, such as describing a scene in a photo using text or creating an image from a descriptive paragraph.

- **Enhanced user experience**: Multimodal LLMs can offer a more immersive and interactive user experience, providing output that engages multiple senses and caters to different learning styles and user preferences.

- **Accessibility improvements**: These capabilities can also enhance accessibility, translating text to speech for the visually impaired or generating sign language animations from speech for the hearing impaired.

- **Richer data interpretation**: By synthesizing information from different modalities, LLMs can provide a more comprehensive interpretation of data, useful in fields such as medical diagnosis, where visual (e.g., MRI scans) and textual (e.g., clinical notes) data must be combined.

- **Creative and design applications**: In creative fields, multimodal LLMs could assist in design processes, providing suggestions for visual content based on textual descriptions or creating drafts for multimedia campaigns.

- **Interactive learning and gaming**: Educational software and games could become more interactive and adaptive with multimodal LLMs, offering learning content and feedback through various forms of media.

The development of multimodal LLMs marks a major advance in AI capabilities. Integrating various data types, these models enable more natural, intuitive interactions, transforming communication, learning, and creation across industries.

Greater personalization

The evolution of LLMs is set to usher in new levels of personalization, enhancing the user experience in a way that feels bespoke and individualized. Let's explore what this might include:

- **User preference learning**: Future LLMs will likely have the ability to learn and remember user preferences over time, adapting their responses and suggestions based on past interactions, user choices, and feedback.

- **Adaptive content delivery**: Content delivery could be dynamically adjusted to match a user's interests, reading level, and engagement patterns. This means that the information presented to the user would be in line with what is most relevant and engaging to them personally.

- **Contextual recommendations**: By analyzing user behavior and context, LLMs could make highly relevant recommendations, similar to how advanced recommendation engines work today, but with a broader and more nuanced understanding of context.

- **Learning style adaptation**: Educational applications of LLMs could adapt to different learning styles, offering explanations, examples, and practice exercises that align with a user's preferred way of learning.

- **Conversational memory**: In conversations, LLMs would be able to recall past discussions, creating a sense of continuity and understanding, much like interacting with a familiar human over time.

- **Emotional intelligence**: Greater personalization also involves emotional intelligence, where an LLM can detect subtle cues in language to understand a user's emotional state and respond in a way that demonstrates empathy.

- **Customized creativity**: In creative tasks, such as writing, designing, or composing music, LLMs could reflect a user's personal style and past creative choices in the content they generate.

- **Personalized language use**: LLMs could adjust the complexity, tone, and type of language used, based on a user's proficiency and comfort with the language, making communication more effective and comfortable.

- **Predictive personalization**: Leveraging predictive analytics, LLMs might anticipate user needs and provide assistance before a user explicitly requests it, based on patterns and inferred intentions.

- **Integrated personalization across platforms**: Personalization could extend across different platforms and devices, providing a consistent and tailored experience, whether the user is on their phone, computer, or using a voice assistant.

- **Ethical and privacy considerations**: As personalization deepens, so does the importance of managing it ethically. Ensuring user privacy and consent will be critical, with LLMs needing to balance personalization with responsible data use.

- **Personalized accessibility features**: Accessibility features could be personalized, with LLMs adjusting the way content is presented to suit individual accessibility needs, such as for users with visual or auditory impairments.

- **Interactive personalized feedback**: For tasks such as learning or fitness, LLMs could provide interactive, personalized feedback that helps users improve their skills or achieve their goals in a way that's most effective for them.

The development of multimodal LLMs significantly advances AI capabilities. By integrating various data types, these models transform communication, learning, and creation across industries.

Increased efficiency and speed

The next generation of LLMs is poised to bring substantial improvements in terms of efficiency and speed. Here are the key areas where these improvements are expected:

- **Model architecture innovations**: Future LLMs will benefit from advancements in neural network architectures, which might include more efficient transformer models or entirely new designs that optimize the way information is processed and retrieved

- **Processing power advancements**: As hardware technology evolves, the increase in processing power will allow LLMs to handle complex computations more quickly, significantly reducing the time required to generate responses

- **Optimized algorithms**: Algorithms used in LLMs will likely become more sophisticated, with better optimization techniques that enable faster data processing without sacrificing the quality of the output

- **Parallel processing techniques**: By leveraging parallel processing, LLMs will be able to handle multiple tasks at once, which will be instrumental in managing a higher number of simultaneous users

- **Quantum computing potential**: Although still in the early stages, quantum computing has the potential to revolutionize the speed at which LLMs operate, by processing vast amounts of data at speeds unattainable with classical computing

- **Distributed computing**: The use of distributed computing, where tasks are spread across multiple machines, will enhance the performance of LLMs, especially in cloud-based services

- **Edge computing integration**: By integrating edge computing, where data processing is done closer to the data source, LLMs will be able to deliver faster responses, particularly in real-time applications

- **Resource-efficient training**: Innovations in training procedures will make LLMs quicker to train, with fewer data requirements and more efficient use of computational resources

- **Cache and memory optimization**: Improvements in caching techniques and memory usage will allow LLMs to retrieve and utilize information more efficiently, speeding up response generation

- **Dynamic scaling**: Cloud services will likely offer dynamic scaling capabilities for LLMs, automatically adjusting the amount of computational power allocated, based on the current demand

- **Energy-efficient computing**: There will be a focus on making computing more energy-efficient, which is not only better for the environment but also allows sustained high-speed processing

- **Real-time interaction capabilities**: For applications that require real-time interaction, such as digital assistants or online gaming, LLMs will be optimized to provide instantaneous responses

- **Streamlined data fetching**: LLMs will become more adept at quickly fetching the necessary information from databases and knowledge bases, making the generation of informed responses much faster

Future LLMs will offer improved user experiences with rapid response times and high-volume request handling, which are crucial, as they integrate into everyday technologies and expand their user base.

Advanced reasoning and problem-solving

Anticipated advancements in the reasoning and problem-solving abilities of next-generation LLMs include the following:

- **Complex decision-making**: Future LLMs will likely be able to navigate complex decision-making scenarios, weighing different factors and potential outcomes to arrive at reasoned conclusions.

- **Enhanced logic processing**: Improvements in logical reasoning will enable LLMs to better understand and apply logical operators and relationships, crucial for technical fields such as programming and mathematics.

- **Abstract reasoning**: Abstract reasoning involves understanding concepts that are not directly observed. LLMs will become better at extrapolating from known information to new, unseen situations or problems.

- **Creativity in problem-solving**: Creativity is not just an artistic trait; it's also key to problem-solving. LLMs will be able to generate novel solutions to problems by combining seemingly unrelated concepts in innovative ways.

- **Strategic planning**: LLMs will improve in planning several steps ahead, a skill necessary for tasks ranging from game playing to strategic business planning.

- **Understanding cause and effect**: Recognizing cause-and-effect relationships will enable LLMs to more accurately predict outcomes and understand the implications of certain actions or events.

- **Cross-domain knowledge application**: Next-generation LLMs will be adept at applying knowledge from one domain to another, using analogical reasoning to solve problems in creative and effective ways.

- **Mathematical reasoning**: Enhanced mathematical reasoning will allow LLMs to perform more complex calculations and provide explanations or solutions to mathematical problems.

- **Ethical reasoning**: As AI becomes more prevalent, the ability of LLMs to reason ethically and consider the moral implications of their suggestions or actions will be crucial.

- **Emotionally intelligent responses**: Emotional intelligence in problem-solving involves understanding human emotions and considering them when proposing solutions, leading to more empathetic and human-centric AI.

- **Scientific reasoning**: LLMs could contribute to scientific research by hypothesizing, designing experiments (virtually), and interpreting data to derive logical conclusions.

- **Interactive learning and feedback**: Interactive learning will allow LLMs to reason through problems by asking questions, receiving feedback, and iteratively refining their understanding and approaches.

- **Dynamic resource allocation**: In computational terms, reasoning and problem-solving often require dynamic resource allocation, which next-generation LLMs could manage effectively to optimize their performance for different tasks.

These advancements will enhance LLM capabilities and expand their utility across sectors such as education, healthcare, finance, and technology, aiming to develop LLMs as advanced cognitive partners in problem-solving tasks.

Broader knowledge and learning

The forthcoming generations of LLMs will likely be distinguished by their expansive knowledge bases and the capacity for real-time learning. Here's what we might expect:

- **Expansive knowledge bases**: LLMs will have access to vast amounts of information, encompassing a wide array of subjects, languages, and cultural contexts, allowing for a much more comprehensive understanding of user queries.

- **Real-time information updating**: Unlike current models that require periodic retraining, future LLMs might continuously update their knowledge base with the latest information, research findings, news, and trends, ensuring that the knowledge they provide is current and relevant.

- **Dynamic learning from interaction**: LLMs will learn from each interaction, refining their models to improve accuracy and relevance for future responses. This could include learning from user corrections, feedback, and engagement metrics.

- **Cross-disciplinary synthesis**: By integrating knowledge from various disciplines, LLMs will be able to synthesize information, providing more nuanced and comprehensive answers that draw from multiple fields of expertise.

- **Personalized knowledge paths**: LLMs will create personalized knowledge paths for users, guiding their learning based on previous interactions, stated goals, and demonstrated interests.

- **Predictive learning**: Anticipating users' needs based on past behaviors, LLMs will preemptively learn and present information that aligns with their predicted queries or tasks.

- **Contextual understanding**: The ability to understand context deeply will allow LLMs to discern which pieces of their broad knowledge are most relevant to present in any given situation.

- **Semantic understanding and reasoning**: Future LLMs will exhibit a more profound semantic understanding, enabling them to reason through complex topics and provide explanations that are not only factually accurate but also contextually meaningful.

- **Collaborative learning**: LLMs may be able to learn collaboratively, both from other AI systems and humans, leveraging the collective intelligence to enhance their knowledge base.

- **Expertise in niche areas**: While having broad knowledge bases, LLMs will also specialize in niche areas, becoming experts in specific domains where depth of knowledge is crucial.

- **Learning from diverse data sources**: Incorporating diverse data sources will prevent knowledge silos and ensure a well-rounded perspective, making LLMs reliable for a wide range of topics.

- **Multilingual and cultural learning**: LLMs will be adept at understanding and learning from multilingual content, enabling them to serve a global user base with cultural awareness and sensitivity.

- **Adaptive learning mechanisms**: LLMs will employ advanced adaptive learning mechanisms to tailor their learning process to the most efficient strategies, based on the task at hand.

These advancements will position LLMs as powerful tools for information dissemination, education, and decision support, making them integral to businesses, educators, researchers, and everyday users.

Ethical and bias mitigation

The integration of ethical considerations and bias mitigation is a pivotal aspect of the ongoing development of LLMs. Here's an expanded view of what this entails:

- **Fairness and equity**: Future LLMs will incorporate algorithms designed to ensure fairness, actively preventing discriminatory outcomes based on race, gender, age, or other sensitive attributes.

- **Bias detection and correction**: Advanced techniques will be developed to detect biases within the training data and model outputs. Once identified, mechanisms will be put in place to correct these biases, ensuring more balanced and fair responses.

- **Privacy preservation**: LLMs will be designed with privacy as a core feature, employing methods such as differential privacy and federated learning to protect user data. They'll handle personal information responsibly, adhering to global privacy standards and regulations.

- **Transparency in decision-making**: There will be an emphasis on explainability, with LLMs providing clear explanations for their decisions and suggestions, enabling users to understand the reasoning behind AI-generated content and actions.

- **Misinformation prevention**: Techniques to recognize and avoid the spread of misinformation will be integral to LLMs. They'll cross-reference facts and check against reputable sources before disseminating information.

- **Ethical training and guidelines**: Ethical training for developers and guidelines for LLMs will be established, ensuring that ethical considerations are embedded in the design and deployment of these models.

- **Inclusive and diverse training data**: To mitigate biases, the training data for LLMs will be curated to be as inclusive and diverse as possible, representing a wide range of voices and perspectives.

- **Cultural sensitivity and localization**: LLMs will be attuned to cultural nuances and local contexts, avoiding generalizations that could lead to insensitive or incorrect responses.

- **User control over data**: Users will have more control over how their data is used by LLMs, including the ability to opt out of data collection or have their data deleted.

- **Robust content moderation**: LLMs will include robust content moderation systems to prevent the generation or amplification of harmful content.

- **Ongoing monitoring and auditing**: Continuous monitoring and regular auditing of LLMs will ensure that ethical standards are maintained and biases are addressed promptly.

- **Stakeholder engagement**: A broader range of stakeholders, including ethicists, sociologists, and representatives from affected communities, will be involved in the development and oversight of LLMs.

- **Legal and ethical compliance**: LLMs will be designed to comply with both the letter and the spirit of laws and ethical norms across jurisdictions, adjusting their behavior to align with local regulations and ethical expectations.

In summary, as LLMs become more sophisticated and widespread, addressing ethical issues and biases is crucial to creating trustworthy, fair, and ethically aligned models that benefit society equitably.

Improved interaction with other AI systems

As we progress toward more integrated AI ecosystems, future LLMs are expected to serve as sophisticated intermediaries between humans and various specialized AI systems. Here's a deeper look into what this might involve:

- **Seamless integration**: LLMs will be designed to seamlessly integrate with a variety of AI systems, enabling smooth data exchange and interoperability without the need for extensive customization.

- **Translation of human requests**: They will be capable of translating complex human requests into the specific formats required by other AI services, acting as a user-friendly interface for more technical or specialized systems.

- **Central coordination role**: LLMs may assume a central coordination role within AI frameworks, directing tasks to the most suitable AI service and ensuring that outputs are combined to provide coherent and comprehensive responses.

- **Inter-AI communication**: Future LLMs will enable different AI systems to communicate with each other, even if they were developed independently, by standardizing the communication protocols and data formats.

- **Contextual relay**: When relaying information between systems, LLMs will provide context to ensure that each AI component understands the relevance of the data it receives or processes.

- **Real-time data synthesis**: LLMs will synthesize real-time data from multiple AI systems to provide up-to-date information and insights, critical for applications such as financial analysis or emergency response.

- **Human-in-the-loop interfaces**: They will facilitate human-in-the-loop interfaces, allowing human operators to step in or review decisions when necessary, thus ensuring that the integration of AI systems remains under human oversight.

- **Dynamic service selection**: By assessing the capabilities and current loads of various AI systems, LLMs will dynamically select the most appropriate service for a given task, optimizing the use of resources.

- **Error handling and diagnostics**: LLMs will assist in identifying and diagnosing errors or inconsistencies across interconnected AI systems, aiding in maintaining system integrity and performance.

- **Automated learning and updating**: They will enable automated learning and updating processes across different AI systems, sharing insights and new data to collectively improve performance.

- **Multi-agent collaboration**: LLMs will facilitate collaboration in multi-agent systems, where several AI agents work together on complex tasks, ensuring that each agent's contributions align with the overall objective.

- **Modular AI development**: The integration capabilities of LLMs will encourage more modular development of AI systems, where individual components can be developed independently but are designed to work together cohesively.

- **Customizable user experiences**: By interacting with different AI services, LLMs will be able to create highly customizable user experiences, combining services in unique ways to meet individual user needs.

In essence, LLMs will evolve into the connective tissue of AI, enabling various services to collaborate more effectively and enhancing AI's capability to perform complex tasks across diverse applications.

More robust data privacy and security

As LLMs become more pervasive in both personal and professional settings, the importance of data privacy and security is paramount. Future developments in this area are expected to include the following:

- **Advanced encryption**: Encryption standards will evolve, with LLMs utilizing more sophisticated encryption methods to secure data both at rest and in transit, ensuring that sensitive information remains confidential

- **Differential privacy**: The implementation of differential privacy techniques will ensure that LLMs can learn from data without compromising the privacy of individuals within a dataset

- **Federated learning**: LLMs may use federated learning to train on decentralized data, allowing a model to learn from user data without that data ever leaving a user's device

- **Secure multi-party computation**: By maintaining the confidentiality of private inputs, **secure multi-party computation** (**SMPC**) facilitates the collaborative computation of a function among multiple entities, thereby enhancing the security of collaborative learning environments

- **Data anonymization and pseudonymization**: Enhanced methods of anonymization and pseudonymization will make it difficult to reverse-engineer personal data from model outputs, or from the model itself

- **Access control and authentication**: Robust access control mechanisms and authentication protocols will be in place to ensure that only authorized individuals can access sensitive data and model functionalities

- **Auditing and compliance**: LLMs will feature comprehensive auditing capabilities to track data usage and access, facilitating compliance with global data protection regulations such as GDPR and CCPA

- **Ethical data use frameworks**: Ethical frameworks for data use will guide the collection, storage, and processing of data within LLMs, ensuring that ethical considerations are at the forefront of data handling practices

- **Decentralized data storage**: Decentralized data storage solutions, such as blockchain, could be employed to enhance security and provide an immutable record of data transactions

- **Consent management**: Improved consent management systems will allow users to have granular control over what data they share and how it is used by LLMs

- **Continuous security monitoring**: LLMs will be monitored continuously for potential security breaches or vulnerabilities, with automated systems in place to detect and respond to threats in real time

- **AI-specific security protocols**: Given the unique nature of AI systems, specialized security protocols will be developed to protect against AI-specific threats, such as model inversion attacks or adversarial inputs

- **Data minimization principles**: Adhering to the principle of data minimization, LLMs will collect and process only the data that is absolutely necessary for the task at hand

These measures will collectively contribute to a more secure and privacy-respecting AI ecosystem. By embedding privacy and security into the fabric of LLMs, we can ensure that these powerful tools can be leveraged safely and responsibly.

Customizable and scalable deployment

The next generation of LLMs is expected to offer a high degree of customization and scalability that will enable businesses of all sizes to tailor AI solutions to their specific needs. Here's what to anticipate:

- **Modular design**: Future LLMs may be designed in a modular fashion, allowing businesses to plug and play different components based on their specific requirements

- **Task-specific customization**: LLMs will be highly customizable, enabling businesses to fine-tune models for specialized tasks, whether it's customer service, data analysis, or content creation

- **Scalability**: These models will be inherently scalable, designed to handle varying loads of work, from small datasets and user bases to vast streams of data and millions of users – without a drop in performance

- **On-demand resources**: Cloud-based deployment will allow for on-demand resource allocation, meaning that businesses can scale their LLM usage up or down as needed, optimizing costs and resources

- **Integration with existing systems**: LLMs will come with tools and APIs that facilitate easy integration with existing business systems and workflows, minimizing the need for extensive overhauls

- **Automated deployment**: Deployment processes will be increasingly automated, using AI itself to assist in the setup and tuning of LLMs for specific business environments

- **Self-optimizing models**: LLMs will have self-optimizing capabilities, continuously learning from their performance and user feedback to improve their accuracy and efficiency over time

- **Industry-specific solutions**: There will be a proliferation of industry-specific LLMs, pre-trained on domain-specific data, which can then be further customized by individual businesses

- **User-friendly interfaces**: Businesses will have access to more user-friendly interfaces for customizing and managing LLMs, reducing the barrier to entry for those without extensive technical expertise

- **Performance monitoring and analytics**: Advanced monitoring and analytics will be built in, providing insights into model performance and helping businesses make data-driven decisions about scaling and customization

- **Edge AI deployment**: Some LLMs will be deployable at the edge, closer to where data is generated, to reduce latency and bandwidth use, particularly important for time-sensitive applications

- **Containerization and microservices**: The use of containerization and microservices architectures will promote more agile deployment and scaling of LLMs

- **Compliance and governance**: Customization options will include compliance controls, ensuring that LLMs adhere to regional regulations and industry standards as they scale

These advancements will make LLMs more accessible and effective across industries, allowing businesses to drive innovation, improve services, and stay competitive with customizable and scalable AI deployment options.

Regulatory compliance and transparency

The future development of LLMs will be heavily influenced by the need for regulatory compliance and transparency. As these models become more integral to various sectors, from healthcare to finance, their alignment with legal and ethical standards becomes critical. Here's what this may entail:

- **Alignment with legal frameworks**: LLMs will be designed to comply with existing and emerging legal frameworks, such as the GDPR in Europe and others globally that regulate data privacy and AI ethics

- **Transparency in AI decision-making**: There will be a greater emphasis on creating LLMs that can explain their decision-making processes in understandable terms, allowing users to grasp how conclusions are reached

- **Audit trails**: LLMs will generate comprehensive audit trails that document the decision-making process, which is essential for compliance purposes and for reviewing AI's actions if there are disputes

- **Bias and fairness assessments**: Regular assessments will be conducted to ensure that LLM outputs are free from bias and that the models operate fairly across different demographics

- **Ethical AI design**: Ethical considerations will be embedded into the design process of LLMs, ensuring that they operate within the accepted moral bounds of society

- **Data handling and consent**: Robust systems to manage user consent regarding data use will be implemented, ensuring that LLMs handle data in a manner that respects user preferences and privacy laws

- **User empowerment**: Users will have more control over what data is used and how it is used by LLMs, including the ability to opt out or correct data

- **Standardization of practices**: Industry-wide standards for the development, deployment, and management of LLMs will likely emerge, ensuring consistency and reliability

- **Risk assessment and management**: LLMs will incorporate mechanisms to assess and manage risks, particularly in areas where AI decisions could have significant impacts on individuals or businesses

- **Interoperability**: LLMs will be designed for interoperability with other systems and AI models, facilitating a seamless exchange of information within a regulated framework

- **Governance structures**: Clear governance structures will be established to oversee LLM operations, including the roles and responsibilities of all parties involved in the AI life cycle

- **Consumer protection**: Measures will be put in place to protect consumers from potential harm caused by LLMs, including incorrect or harmful content generation

- **Open standards and protocols**: The use of open standards and protocols will likely be encouraged to promote transparency and allow independent verification of LLM compliance and performance

- **Cross-border compliance**: LLMs will need to navigate the complexities of cross-border compliance, adhering to the laws and regulations of all the jurisdictions they operate in

By proactively addressing regulatory compliance and transparency, developers and users of LLMs can foster trust in AI technologies. These measures are not only about adhering to regulations but also about ensuring that LLMs are used responsibly and ethically, contributing positively to society and instilling confidence in their output.

Accessible AI for smaller businesses

The trend toward making powerful LLMs more accessible to smaller businesses marks a significant democratization of AI technology. Here's an overview of how this could unfold:

- **Cost-effective solutions**: Innovations in AI will lead to more cost-effective LLM solutions, reducing the financial barriers that often prevent smaller businesses from leveraging advanced AI technologies

- **Simplified integration**: LLM providers will likely offer simplified integration options, with Plug and Play solutions that can be easily incorporated into existing business processes without the need for specialized expertise

- **Cloud-based services**: Cloud-based AI services will enable small businesses to use state-of-the-art LLMs without the need for significant hardware investments, paying only for the services they use

- **User-friendly platforms**: The rise of user-friendly AI platforms, with intuitive interfaces and guided workflows, will allow smaller businesses to implement and manage LLMs without the need for in-house AI specialists

- **Pre-trained models**: Smaller businesses will have access to pre-trained models that can be fine-tuned for their specific needs, avoiding the costs and complexities of training models from scratch

- **Scalable performance**: LLMs will be scalable in performance, ensuring that small businesses can start with modest AI implementations and scale up as their needs grow

- **Tailored business applications**: AI developers will create tailored applications, designed for the unique challenges and opportunities of small businesses across various industries

- **Educational resources and support**: An increase in educational resources and community support will empower smaller businesses to make informed decisions about AI and how to implement LLMs effectively

- **Subscription models**: Subscription-based models will provide smaller businesses, with the flexibility to use advanced LLMs without upfront capital investments

- **Marketplaces for AI services**: Online marketplaces for AI services will emerge, where businesses can find and deploy LLMs suited to their specific tasks and industries

- **API economy**: The expansion of the API economy will give smaller businesses the ability to integrate LLMs into their operations through simple API calls

- **Regulatory support**: Regulations may evolve to support smaller businesses in adopting AI, possibly through incentives or frameworks that lower the entry barrier

- **Community-driven development**: Open source projects and community-driven AI development will provide small businesses with access to high-quality, collaboratively created LLMs

These advancements will not only make AI more affordable and accessible but also enable smaller businesses to compete more effectively with larger corporations, fostering innovation and growth across the entire business landscape.

Enhanced interdisciplinary applications

LLMs are set to become even more versatile, with enhanced applications across a broad range of fields. The interdisciplinary use of LLMs will be driven by their ability to understand and generate domain-specific content, analyze complex data, and interact with users in contextually relevant ways. Here's what we can expect in various sectors:

- **Healthcare**: In healthcare, LLMs will be able to digest medical literature and patient data to assist with diagnosis, treatment planning, and patient education. They could help to parse complex medical records, providing summaries for healthcare providers, and even engage in patient monitoring by analyzing notes and reports for signs of change in a patient's condition.

- **Education**: LLMs will revolutionize education by offering personalized learning experiences, automating administrative tasks, and providing tutoring in a range of subjects. They could adapt to individual students' learning styles and progress, recommend resources, and assess student work.

- **Legal industry**: In the legal field, LLMs will assist in researching case law, drafting documents, and even predicting litigation outcomes. They could provide support in understanding complex legal language for both professionals and the public.

- **Creative industries**: For creative industries, LLMs will aid in content creation, from writing scripts to generating artistic concepts. They could also function as design assistants, proposing ideas and refining creative works based on user input.

- **Customer service**: Customer service will be bolstered by LLMs capable of conducting sophisticated conversations, handling inquiries, and resolving issues, with a level of personalization and understanding that mirrors human agents.

- **Financial services**: In financial services, LLMs will analyze market reports, financial statements, and economic data to provide insights, forecast trends, and personalize financial advice for clients.

- **Scientific research**: LLMs will assist researchers by sifting through vast amounts of scientific publications to identify relevant studies, generate hypotheses, or even draft research papers.

- **Engineering**: Engineers could use LLMs to interpret technical specifications, generate CAD drawings from descriptions, or simulate how changes in design could affect performance.

- **Supply chain and logistics**: LLMs will optimize supply chain operations by predicting disruptions, automating communications, and providing real-time analysis of logistics data.

- **Environmental sciences**: LLMs could process environmental data to model climate change effects, suggest conservation strategies, or generate reports on biodiversity.

- **Public sector**: Governments and public sector organizations will employ LLMs to enhance citizen services, draft policy, and analyze public feedback for better governance.

- **Language translation and localization**: LLMs will provide advanced translation services and localization, making content accessible across linguistic and cultural boundaries with a high degree of accuracy.

- **Psychology and mental health**: In mental health, LLMs could support therapy sessions by providing conversational agents that help with stress relief, or they could analyze patient language to support diagnosis and treatment.

- **Agriculture**: LLMs will aid in agricultural planning and management by analyzing reports and data, providing farmers with actionable insights.

As LLMs advance, their interdisciplinary applications will drive innovation and efficiency across sectors, transforming professional practices by processing and generating specialized knowledge.

The evolution of LLMs represents a convergence of technological progress, user-centric design, and ethical AI governance. As these models become more advanced, they will present both new opportunities and challenges that will shape the future of human-AI interaction.

Getting ready for GPT-5 – infrastructure and skillsets

Preparing for the arrival of GPT-5, or any advanced iteration of LLMs, involves several key areas of focus for businesses and individuals looking to leverage the technology. Here's an outline of what preparation could involve:

- **Infrastructure readiness**:

  - **Cloud services**: Ensuring access to scalable cloud services that can support the high computational demands of GPT-5

  - **Data storage**: Upgrading data storage solutions to handle the increased volume of data processing and interactions

  - **Security measures**: Implementing robust cybersecurity measures to protect the data that GPT-5 will process and secure AI's output

  - **High-speed connectivity**: Investing in high-speed internet connections to facilitate real-time interactions with cloud-based GPT-5 services

  - **API integration**: Developing or updating APIs for the smooth integration of GPT-5 capabilities into existing systems and applications.

  - **Hardware accelerators**: Utilizing hardware accelerators, such as GPUs or TPUs, to locally run intensive machine learning tasks where latency is a critical factor

- **Skillset enhancement**:

  - **AI literacy**: Building AI literacy across an organization to ensure that all levels of staff understand the capabilities and limitations of GPT-5

  - **Technical training**: Providing technical training for IT teams on managing and maintaining AI systems, including GPT-5

  - **Data science skills**: Investing in data science skills, including understanding how to work with large datasets, model training, and fine-tuning

  - **AI ethics and governance**: Understanding the ethical considerations and governance required when deploying AI at scale, including bias mitigation and data privacy

  - **Change management**: Preparing a workforce for change management, ensuring that the introduction of GT-5 enhances workflows without causing disruption

  - **Creative problem-solving**: Encouraging skills in creative problem-solving and design thinking to fully utilize the generative aspects of GPT-5

  - **Advanced prompt engineering**: Developing expertise in fine-tuning GPT-5 to deliver more relevant, accurate, and context-aware responses, aligned with business needs

  - **Interdisciplinary collaboration**: Fostering a culture of interdisciplinary collaboration, as GPT-5's applications will span across various departments and industries

- **Organizational strategies**:

  - **AI-first approach**: Adopting an AI-first approach in strategic planning, considering how GPT-5 can be used to achieve business goals

  - **Innovation labs**: Establishing innovation labs or task forces to explore and experiment with GPT-5 applications relevant to the business

  - **Partnerships**: Forming partnerships with AI research institutions and technology providers to stay at the forefront of AI developments

  - **Pilot projects**: Running pilot projects to understand the impact of GPT-5 and identify best practices for wider deployment

  - **Feedback mechanisms**: Creating feedback mechanisms to continuously learn from GPT-5 interactions and improve user experience and outcomes

  - **Prepare for higher compute requirements**: Plan for the increased computational demands of deploying GPT-5 by investing in scalable cloud solutions, high-performance computing, and efficient data storage to support large-scale AI workloads

Preparing for GPT-5 is not just about technology and skills; it's also about cultivating a forward-thinking culture that is ready to embrace the transformative potential of AI while addressing the challenges and responsibilities it brings.

Potential breakthroughs and challenges ahead

As the field of AI and, more specifically, LLMs continues to evolve, we are likely to witness several breakthroughs along with significant challenges. Here's a look at what the future may hold:

- **Potential breakthroughs**:

  - **Advanced cognitive understanding**: LLMs may develop a deeper understanding of context, sarcasm, and nuanced language, effectively managing tasks that require a high level of cognitive abilities

  - **Multimodal capabilities**: The ability to process and generate multimodal content, integrating text with images, audio, and video, could revolutionize how we interact with AI

  - **Personalized AI interactions**: Breakthroughs in personalization could lead to LLMs that adapt to individual user's preferences, learning styles, and needs in real time

  - **Generalized AI**: We could move toward more generalized AI that can perform a variety of tasks without extensive retraining or fine-tuning

  - **Quantum computing integration**: Integration with quantum computing may dramatically increase the speed and capacity of LLMs, enabling them to solve complex problems previously deemed unsolvable

  - **Language and culture translation**: LLMs could become powerful tools for breaking down language and cultural barriers, providing accurate and context-aware translations in real time

  - **AI-assisted research and development**: In fields such as pharmaceuticals and environmental science, LLMs might significantly speed up research and development cycles

  - **Ethical AI governance**: The establishment of robust ethical AI governance frameworks could ensure that LLMs are developed and used responsibly

- **Challenges ahead**:

  - **Bias and fairness**: Despite improvements, ensuring that LLMs are free from bias and treat all users fairly remains a major challenge

  - **Data privacy**: Balancing the data needs of LLMs with the privacy rights of individuals will continue to be a complex issue, especially with varying global regulations

  - **Explainability**: As LLMs become more complex, making their decision-making processes transparent and understandable to non-experts is a significant challenge

  - **Misinformation control**: Preventing LLMs from generating or propagating misinformation requires sophisticated understanding and filtering mechanisms

  - **Security threats**: The risk of malicious use of LLMs, such as creating deepfakes or automated hacking tools, is a concern for cybersecurity

- **Resource intensity**: The environmental impact of training and running large-scale LLMs, which require significant computational resources, is a growing concern

- **Intellectual property questions**: The ability of LLMs to generate content raises complex questions about copyright and intellectual property rights

- **Dependency and skill erosion**: Over-reliance on LLMs could lead to the erosion of human skills in areas such as writing, analysis, and decision-making

- **Regulatory compliance**: As AI regulations are developed and become more stringent, ensuring that LLMs comply with these regulations is a challenge

- **Interdisciplinary integration**: Effectively integrating LLMs into interdisciplinary fields, each with its own complexities and nuances, requires extensive domain-specific expertise

- **AI ethics**: Developing AI in an ethical way, particularly as it becomes more autonomous and capable, is a profound challenge

- **Access and equity**: Ensuring equitable access to the benefits of LLMs across different socioeconomic, geographic, and cultural groups remains a challenge

The road ahead for LLMs is filled with both exciting possibilities and formidable obstacles. The true potential of LLMs will be realized through a collaborative effort, involving technologists, ethicists, policymakers, and the broader community, who address these challenges and guide the development of these powerful systems for the greater good.

Strategic planning for future LLMs

Strategic planning to integrate and utilize future LLMs in businesses and organizations involves several key steps and considerations to ensure that these advanced tools are harnessed effectively and responsibly. Here's an outline of what such strategic planning might involve:

- **Assessment of organizational needs and goals**:

  - **Identify opportunities**: Determine where LLMs can solve existing problems or create new opportunities

  - **Set objectives**: Define clear objectives for what an organization aims to achieve with LLMs

- **Resource allocation**:

  - **Budgeting**: Allocate budget for infrastructure, training, and ongoing costs associated with LLM deployment

  - **Talent acquisition**: Invest in hiring or training personnel with the expertise to manage and work with LLMs

- **Infrastructure preparation**:

  - **Technology investment**: Upgrade existing infrastructure to support the computational demands of LLMs

  - **Data management**: Establish robust data management practices to feed accurate data to the LLMs

- **Risk management**:

  - **Ethical considerations**: Plan for the ethical implications of using LLMs, including biases and decision-making impacts

  - **Data privacy**: Ensure that the use of LLMs complies with data privacy laws and regulations

- **Compliance and legal check**:

  - **Regulatory review**: Stay abreast of AI regulations that could affect the use of LLMs

  - **Intellectual property**: Address intellectual property concerns related to generated content

- **Technology partnerships**:

  - **Collaborate with AI leaders**: Partner with tech firms and AI research institutions for access to the latest developments

  - **Ecosystem engagement**: Engage with the broader AI ecosystem, including start-ups and academic entities

- **Employee training and development**:

  - **Upskilling programs**: Implement training programs to upskill employees in AI literacy

  - **Change management**: Prepare your workforce for changes in workflow and processes due to AI integration

- **Pilot testing**:

  - **Proof of concept**: Start with a proof of concept to test the value and integration of LLMs within your organization

  - **Iterative approach**: Use an iterative approach to gradually expand the scope of LLM applications, based on initial learnings

- **Knowledge management**:

  - **Documentation**: Keep thorough documentation of how LLMs are used and the knowledge they generate

  - **Knowledge sharing**: Facilitate knowledge sharing about LLM capabilities and best practices within your organization

- **Monitoring and evaluation**:

  - **Performance metrics**: Establish metrics to evaluate the performance and impact of LLMs

  - **Feedback loops**: Create mechanisms to gather feedback from users and adjust strategies accordingly

- **Future-proofing**:

  - **Scalability**: Ensure that LLM solutions are scalable to grow with your organization

  - **Flexibility**: Maintain flexibility in AI strategies to adapt to the rapidly evolving AI landscape

- **Ethical AI framework**:

  - **Develop ethical frameworks**: Create guidelines to ensure that AI is used ethically throughout your organization

  - **Transparency**: Plan for transparent AI operations, where stakeholders understand how and why AI makes decisions

- **Long-term vision**:

  - **Strategic AI vision**: Develop a long-term vision for how LLMs can transform the organization

  - **Innovation culture**: Foster a culture of innovation that embraces AI as a tool for enhancement, not replacement

Strategic planning for LLMs is an ongoing process that must be revisited regularly as technology evolves. It requires a multidisciplinary approach, involving stakeholders from various departments, to ensure that LLMs are implemented in a way that aligns with an organization's values and objectives.

Summary

Future LLMs, including GPT-5, will offer advanced capabilities in understanding, contextualization, and multimodal processing, improving user experience and expanding applications. They will provide greater personalization, efficiency, and speed, making interactions more natural and adaptive.

Enhanced reasoning and problem-solving abilities will position LLMs as cognitive partners in various fields. Broader knowledge bases and real-time learning will transform information dissemination, education, and decision support. Ethical considerations and bias mitigation will be prioritized, ensuring fairness, privacy, and transparency. Improved interaction with other AI systems will enable seamless integration and enhanced functionality across applications.

Preparing for GPT-5 requires upgrading infrastructure (e.g., cloud services, data storage, security, and connectivity) and enhancing skillsets (e.g., AI literacy, technical training, data science, and ethics). Strategic planning involves adopting an AI-first approach, setting up innovation labs, forming partnerships, and running pilot projects. Addressing breakthroughs such as advanced understanding and multimodal capabilities, while managing challenges in bias, data privacy, and ethical AI use, ensures effective and responsible deployment.

In the next and final chapter, we will conclude our book.

15

Conclusion and Looking Forward

Bringing the book to a close, we will synthesize the key insights gained throughout the reading journey. We will offer a forward-looking perspective on the trajectory of LLMs, pointing you toward resources for continued education and adaptation in the evolving landscape of AI and NLP. The final note will encourage you to embrace the LLM revolution with an informed and strategic mindset.

In this chapter, we're going to cover the following main topics:

- Key takeaways from the book
- Continuing education and resources for technical leaders
- Final thoughts – embracing the LLM revolution

By the end of this chapter, you should have a consolidated understanding of the fundamental concepts and strategic insights provided throughout this book and an understanding of ongoing learning and professional growth.

Key takeaways from the book

A comprehensive exploration of LLMs spans a wide array of topics, from their foundational architecture to cutting-edge strategies for their deployment and optimization. The key takeaways of the various facets involved in the development and application of LLMs offered in this guide are discussed next.

Foundational architecture and decision-making

The foundational architecture of LLMs is a rich tapestry of interconnected systems and algorithms that give them the ability to process and understand human language with remarkable proficiency. *Chapter 1, LLM Architecture*, posited that a deep comprehension of the "anatomy" of LLMs is essential for any meaningful discourse on their capabilities and limitations. This anatomy refers to the various structural and functional components that make up an LLM. Central to this discussion was the concept of transformer models and attention mechanisms, which are the building blocks of the most advanced LLMs to date.

Transformer models represent a significant departure from earlier architectures such as RNNs. Unlike RNNs, which process input data sequentially and may struggle with long-range dependencies in text, transformers employ a mechanism that allows them to weigh the importance—or "attention"—of different parts of the input data, regardless of their position. This means that transformers can effectively understand the context and nuanced relationships within a sentence or across multiple sentences, which is vital for tasks such as translation, question-answering, and summarization.

The model's ability to predict each part of the output sequence is greatly enhanced by the attention mechanism—a core feature of transformers—enabling it to focus on different sections of the input sequence. This function mirrors the way a human reader might revisit a sentence to gain better comprehension. This allows for a dynamic allocation of computational resources to parts of the input where they are most needed, enhancing the model's ability to understand and generate language.

Building on the architectural foundation, *Chapter 2, How LLMs Make Decisions*, explored how LLMs make decisions. The decision-making process of an LLM is not a straightforward execution of hardcoded instructions; rather, it is a complex interplay of probabilities and statistical patterns learned from data. When an LLM generates a response, it is essentially calculating the probability of a sequence of words given its training data and the input it has received. This involves statistical models that can interpret intricate patterns in text data to produce relevant and coherent language outputs.

However, this decision-making prowess is not without its challenges. *Chapter 2* acknowledged that biases inherent in the training data can skew the decisions made by LLMs, leading to outputs that may perpetuate stereotypes or inaccuracies. Additionally, reliability issues arise because LLMs, like all statistical models, are susceptible to errors, particularly when faced with ambiguous or novel input that falls outside their training data.

The landscape of LLM decision-making is thus one of constant evolution. Researchers and practitioners in the field are continually seeking ways to refine the decision-making process. This involves not only tweaking the architecture and the training data but also developing new methodologies to address biases and improve the reliability of these models. Future advancements in LLMs will likely focus on creating more robust, fair, and interpretable models that can make decisions in a manner that is both transparent and aligned with ethical standards.

Training mechanics and advanced strategies

The mechanics of training LLMs constitute a cornerstone of their successful application and functionality. *Chapter 3, The Mechanics of Training LLMs*, elaborated on the intricate processes involved in preparing these sophisticated models to perform a vast array of linguistic tasks. The robust performance of LLMs is contingent upon the meticulous preparation and management of data, which is the raw material from which these models learn. The quality, diversity, and representativeness of training data directly influence the model's ability to generalize knowledge to new, unseen examples.

Data preparation goes beyond mere collection; it involves cleaning, labeling, and sometimes augmenting the data to ensure that it can effectively train the model. This also means that the data must be free from errors, well-structured, and inclusive of the various nuances of natural language. Effective management of this data is equally essential. It involves organizing and storing the data efficiently, ensuring that it can be accessed and processed by the LLM without introducing latency or bottlenecks in the training process.

Hyperparameters (the settings that govern the training process) are another focal point of training LLMs. These include learning rate, batch size, and the number of layers in the neural network. Correct hyperparameter settings are crucial; improper adjustments can cause the model to underfit—too simplistic to perform well even on the training data—or overfit, performing well on training data but poorly on new examples.

Chapter 4, Advanced Training Strategies, built on the foundational training mechanics by discussing advanced training strategies that further enhance the performance of LLMs. Transfer learning is one such strategy, allowing models to apply knowledge learned from one task to improve performance on another related task. This is particularly useful when training data for a specific task is scarce, as it enables the model to leverage larger datasets from related tasks to improve its understanding.

Curriculum learning is another advanced strategy where the model is gradually introduced to more complex tasks, much like a human learner progressing from simple to complex concepts. This approach helps in better generalization and often results in a more robust model. Multitasking, where the model is trained on several tasks simultaneously, can also enhance performance by encouraging the model to develop representations that are useful across different linguistic tasks.

These chapters emphasized the importance of a nuanced approach to training LLMs, suggesting that there is no one-size-fits-all solution. Different applications may require different data, hyperparameters, and training strategies. Understanding these methodologies in depth enables practitioners to fine-tune LLMs to their specific needs, leading to more effective, efficient, and reliable models that can perform a wide range of NLP tasks with greater accuracy. This nuanced understanding is critical as the field of NLP continues to evolve, demanding ever more sophisticated and specialized LLMs for an expanding array of applications.

Fine-tuning, testing, and deployment

Fine-tuning LLMs for specific applications is a critical process that ensures these models are not just jacks-of-all-trades but also masters of the specific tasks they are deployed for. *Chapter 5, Fine-Tuning LLMs for Specific Applications*, detailed the intricacies of this customization process, which involves adjusting and adapting pretrained models to perform optimally on tasks such as powering chatbots, translating languages, and conducting sentiment analysis. This fine-tuning process is essential for achieving a level of nuanced understanding and interaction in the model's language output that closely aligns with human-like comprehension and responsiveness.

Tailoring LLMs requires a deep understanding of the domain in question. For instance, a chatbot designed for customer service must understand and generate conversational language that is polite, empathetic, and informative. In contrast, sentiment analysis requires the model to detect subtle cues in text that may indicate positive, neutral, or negative sentiments, which involves training on data annotated with emotional valence. Language translation LLMs must grasp the nuances of grammar, idioms, and cultural context across different languages. This tailoring is achieved through targeted datasets, specific to the task, and training the model further on this data—a process that refines its capabilities and sharpens its focus on the task-specific features.

Testing and evaluation, as elaborated in *Chapter 6, Testing and Evaluating LLMs*, go beyond mere performance metrics such as accuracy or speed. They encompass a range of quantitative metrics and qualitative methods to assess how well the model performs in real-world scenarios. Human-in-the-loop evaluation is particularly emphasized, where human evaluators assess the model's outputs to ensure they meet the required standards of quality, relevance, and appropriateness. This step is also crucial in addressing biases that may be present in the model due to skewed training data or unintentional prejudices encoded in the algorithms. By including human judgment in the loop, ethical and equitable assessments of LLMs can be conducted, ensuring that the models act in a manner that is aligned with societal values and norms.

Chapter 7, Deploying LLMs in Production, discussed the deployment of LLMs in production environments, which requires careful strategic planning. Scalability is a primary consideration, ensuring that the LLM can handle the expected load and perform efficiently at scale. Security best practices are also paramount, as LLMs often process sensitive data that must be protected from unauthorized access and breaches. The deployment phase also involves setting up continuous monitoring and maintenance routines to ensure that the models do not degrade over time or start to produce errors, maintaining their reliability and efficiency long after they have been integrated into production systems.

Integrating LLMs into existing systems, as covered in *Chapter 8, Strategies for Integrating LLMs*, is a non-trivial task that demands a thorough evaluation of compatibility. The integration techniques must be seamless, causing minimal disruption to existing workflows and systems. This requires meticulous planning and testing to ensure that the integration is smooth and that the LLMs can communicate effectively with other components of the system. Security measures are again underscored here, as the integration introduces new vectors through which vulnerabilities could be exploited. Ensuring data integrity and maintaining the trust of users and stakeholders in integrated systems are of utmost importance, as they are the foundation upon which the practical utility of LLMs rests.

In essence, these chapters provided a holistic view of the journey an LLM takes from being a general-purpose model to a specialized tool tailored for specific applications, evaluated rigorously for performance and fairness, and deployed strategically into production environments where it can provide value while operating securely and efficiently.

Optimization, vulnerabilities, and future prospects

The nuanced process of optimizing LLMs for performance is a multifaceted endeavor that incorporates a variety of techniques, as discussed in *Chapter 9, Optimization Techniques for Performance*, and *Chapter 10, Advanced Optimization and Efficiency*. These techniques are not merely about enhancing the computational efficiency of LLMs but are also about making them viable and sustainable for large-scale deployment.

Quantization is one such technique that reduces the precision of the model's parameters, thereby decreasing the model's size and speeding up inference without significantly affecting its performance. By using lower-precision numerical formats, quantization ensures that LLMs can operate on less powerful devices and with lower latency, which is critical for applications that require real-time processing.

Knowledge distillation is another technique where a smaller, more compact model—often referred to as the "student"—is trained to emulate the behavior of a larger, pretrained model—the "teacher." This process allows the distilled model to retain much of the performance of the larger model while being more efficient to run. Knowledge distillation is particularly useful when deploying LLMs in environments with strict resource constraints.

Hardware acceleration, involving the use of specialized hardware such as GPUs or TPUs, is critical for the training and inference processes of LLMs. These hardware solutions are designed to handle the parallelizable nature of deep learning computations, offering substantial improvements in speed and efficiency.

Optimizing data representation goes hand in hand with these techniques. It involves encoding data in a format that maximizes the model's ability to learn from it efficiently. This might involve techniques such as tokenization, vectorization, and the use of embeddings to represent text in a form that captures semantic meaning while being computationally tractable.

Chapter 11, LLM Vulnerabilities, Biases, and Legal Implications, then transitioned into a critical discussion on the vulnerabilities, biases, and legal implications of LLMs. It recognized that while LLMs are powerful tools, they are not immune to the imperfections of their training data or the intentions of their developers. Biases, whether intentional or unintentional, can manifest in outputs, and vulnerabilities can be exploited, potentially leading to unethical uses or outcomes. Thus, the chapter emphasized the necessity of ethical decision-making in the development and application of LLMs and adherence to regulatory frameworks to protect individuals' rights and ensure the responsible use of technology.

The subsequent chapters (*Chapter 12* through *Chapter 14*) took a more application-oriented and forward-looking approach. *Chapter 12*, *Case Studies – Business Applications and ROI*, examined the business applications of LLMs, delving into how they can be implemented to drive ROI. By presenting real-world case studies, this chapter illustrated the tangible benefits that LLMs bring to various industries, from enhancing customer service with intelligent chatbots to automating content creation and analysis tasks.

Chapter 13, *The Ecosystem of LLM Tools and Frameworks*, surveyed the landscape of tools and frameworks available for LLM development, drawing a comparison between open source and proprietary options. The choice between these two types of tools can significantly impact the development process, cost, and innovation. Open source tools often encourage community collaboration and innovation, while proprietary tools might offer specialized functionalities with commercial support.

Looking to the future, *Chapter 14*, *Preparing for GPT-5 and Beyond*, prepared you for the next wave of advancements in LLM technology, such as the anticipated GPT-5. It underscores the importance of strategic planning and the need for businesses and developers to remain adaptable to integrate these advancements effectively. The continuous evolution of LLMs necessitates a proactive approach to infrastructure and skills development, ensuring that practitioners are well-equipped to leverage the capabilities of future models.

In conclusion, the insights provided throughout the chapters present a detailed picture of the transformative influence of LLMs in the field of language processing and interaction. The journey from the foundational aspects of LLMs to their deployment and the anticipation of future advancements highlights the dynamic and ever-progressing nature of AI and NLP. For those in the field, a commitment to continuous education and adaptation is indispensable for fostering innovation and staying ahead in a rapidly evolving technological landscape. The comprehensive overview suggests that, as LLMs grow in sophistication, the ethical, practical, and strategic considerations surrounding them will become increasingly significant, shaping the way we interact with and benefit from these powerful tools.

Continuing education and resources for technical leaders

For technical leaders looking to stay informed and educated about LLMs in the ever-evolving field of NLP, continuing education is crucial. There are numerous resources and avenues available:

- **Online courses and specializations**: Platforms such as Coursera, edX, and Udacity offer courses and specializations that focus on AI, ML, and NLP. These often include modules on LLMs, covering foundational principles, the latest research, and practical implementations.

- **Workshops and conferences**: Attending workshops and conferences such as NeurIPS, ICML, ACL, and others dedicated to AI and ML is a great way for leaders to gain insights into the latest advancements in LLMs and network with peers and experts in the field.

- **Academic journals and publications**: Staying abreast of research through journals such as *Journal of Machine Learning Research, Transactions of the Association for Computational Linguistics*, or *Natural Language Engineering* can provide deep dives into current studies and developments.

- **Professional development programs**: Many universities and institutes offer executive education or professional development programs tailored for leaders in tech. These can be short-term courses that provide overviews of trends and strategic applications of LLMs.

- **Webinars and online tutorials**: Experts in the field often host webinars or create online tutorials discussing the nuances of LLMs. These can be found on platforms such as YouTube or through professional networks such as LinkedIn.

- **Collaborative research initiatives**: Participating in or sponsoring collaborative research with academic institutions can provide first-hand experience with cutting-edge LLM research and development.

- **Technical meetups and peer groups**: Joining local or virtual meetups and peer groups focused on AI and ML can facilitate knowledge sharing and problem-solving with contemporaries.

- **Vendor-specific training**: Companies that provide cloud AI services, such as Google, Amazon, and Microsoft, also offer training and certifications for their specific tools and platforms, which often include modules on LLMs.

- **Books**: There are many comprehensive books on ML and NLP that include sections on LLMs. For more current content, e-books and online publications are updated more frequently than traditional textbooks.

- **MOOCs for specific tools**: For technical leaders interested in specific tools such as TensorFlow, PyTorch, or GPT, MOOCs offer specialized courses that focus on the practical aspects of implementing LLMs using these tools.

- **Internal training sessions**: Organizing regular training sessions within the company led by internal or external experts can help keep the entire tech team updated on LLMs.

- **Mentorship programs**: Establishing mentorship relationships with knowledgeable individuals in the field can provide personalized guidance and learning opportunities.

For technical leaders, it's important to combine these resources with strategic thinking about how LLMs can be applied within their specific business context. This requires not only a technical understanding of the models but also an awareness of the ethical, social, and business implications of their deployment. Regularly updating their knowledge and skills in this domain is key to maintaining a competitive edge and driving innovation within their organizations.

Final thoughts — embracing the LLM revolution

As we stand at the start of the LLM revolution, it is clear that these advanced AI systems are rapidly reshaping the landscapes of numerous industries, from technology to healthcare, and beyond. The transformative power of LLMs to understand, interpret, and generate human language with a previously unattainable depth and nuance marks a paradigm shift not only in ML but also in the very interface between humanity and digital technology.

The integration of LLMs within the business sector represents a transformation that is both deep and wide-ranging. LLMs have transcended the boundaries of mere potential and have become active, influential players in the corporate world. They are not merely auxiliary tools but have become central to strategic business operations, deeply embedded in the decision-making processes, customer service protocols, and marketing strategies.

LLMs have ushered in a new era of customer engagement and service personalization that aligns closely with what was once envisioned in speculative fiction. These models possess the remarkable ability to analyze extensive volumes of text data, which include customer feedback, interaction logs from support services, and the constant stream of conversations happening on social media platforms. This capability allows businesses to extract meaningful patterns, sentiments, and preferences from unstructured data, providing a window into the collective mind and mood of their consumer base.

The implications for business innovation are vast. With the insights gained from LLM analysis, companies can tailor their product development to meet the nuanced needs and desires of their customers. They can identify gaps in the market, anticipate trends, and respond to consumer feedback with agility and precision. This feedback loop can drive continuous product refinement and innovation, ensuring that offerings remain relevant and competitive.

In marketing, LLMs are transforming how businesses connect with their customers. Marketing campaigns can be highly targeted and personalized, delivering messages that resonate on an individual level. This is achieved through the analysis of language patterns, which allows businesses to understand the motivations behind customer behaviors and preferences. By doing so, businesses can craft campaigns that speak directly to the interests and needs of their audience, resulting in more effective marketing efforts.

Customer service has also been revolutionized by the deployment of LLMs. With the capacity to understand and respond to customer queries in natural language, LLMs are powering chatbots and virtual assistants that provide instant, 24/7 support. Immediate assistance enhances the customer experience and lightens the workload for human customer service representatives, allowing them to address more complex queries and issues.

The integration of LLMs in business is creating a virtuous cycle of engagement and improvement. The insights gained from customer interactions lead to better products and services, which, in turn, lead to happier customers who are more engaged and likely to provide further feedback. This cycle is a powerful engine for continuous growth and improvement.

However, the use of LLMs in business also necessitates a commitment to responsible AI practices. Businesses must ensure that the insights gained through LLMs are used ethically, respecting customer privacy and data protection laws. Furthermore, the need to address and mitigate any biases within LLMs is crucial to ensure that the services and products developed do not inadvertently perpetuate inequalities.

The revolution heralded by LLMs is redefining operational efficiency across numerous sectors. The automation of labor-intensive tasks such as document analysis, report generation, and the crafting of complex legal and technical materials is a testament to the advanced capabilities of these AI systems. LLMs, with their ability to quickly process and generate precise and contextually relevant text, are not just streamlining workflows but are also reshaping the very nature of work itself.

Historically, the tasks now being relegated to LLMs required extensive human effort, expertise, and time. By assuming responsibility for these functions, LLMs are liberating human workers from the tedium of routine, repetitive tasks. Document analysis, which involves the review and synthesis of large volumes of information, can be exponentially accelerated with LLMs, which can parse through thousands of pages of text, extract key information, and present summaries that would take humans much longer to produce.

Similarly, the generation of reports (a staple activity in many business operations) can benefit from the deployment of LLMs. By automating the aggregation of data and its conversion into coherent narratives, LLMs enable a level of responsiveness and productivity that was previously unattainable. Furthermore, in the legal and technical domains, drafting documents is a high-stakes task that requires precision and a deep understanding of specific terminologies. LLMs are increasingly able to produce first drafts or assist significantly in the creation of such materials, adhering to the required formalities and specifications.

The reallocation of human cognitive resources away from these tasks opens up new horizons for workforce engagement. With LLMs handling the brunt of data processing and text generation, human workers can redirect their focus toward activities that inherently require human ingenuity, empathy, and strategic thinking. These include creative endeavors, complex problem-solving, and strategic planning, which are areas where human talent truly excels and machines have yet to make significant inroads.

Moreover, this shift has the potential to enhance job satisfaction and personal fulfillment for workers. Engaging in more dynamic and creative work can lead to a more stimulated and motivated workforce. It also allows employees to develop and utilize a broader range of skills, which can lead to increased job satisfaction and opportunities for professional growth. By being freed from routine tasks, workers can focus on building relationships, brainstorming innovative ideas, and contributing to the strategic direction of their organizations.

However, it is essential to approach this transition with a nuanced understanding of the potential disruptions it may cause. There are concerns about job displacement and the need to ensure that the workforce is adequately trained and reskilled to thrive in an AI-augmented environment. Organizations and policymakers must work together to manage the transition, providing educational and training opportunities to equip workers with the skills needed in a changing labor market.

The educational sector stands on the brink of a significant transformation with the advent of LLMs. The traditional one-size-fits-all approach to education is being challenged by the potential of LLMs to provide bespoke learning experiences that cater to the unique needs of each student.

LLMs, with their advanced processing capabilities, can function as AI tutors, capable of assessing and adapting to an individual's learning style, pace, and interests. This adaptive learning can tailor educational content and delivery to the student's current level of understanding, providing personalized explanations and clarifying complex concepts in a manner that is most effective for the individual learner. These AI tutors can interact with students in real time, responding to their queries, guiding their thought processes, and providing immediate, targeted feedback.

Moreover, LLMs can assist in curriculum development by suggesting resources and designing learning activities that align with the student's learning trajectory. This not only enhances engagement by presenting materials in a way that is interesting and relevant to the student but also fosters a deeper understanding by connecting new information to existing knowledge in meaningful ways.

In higher education and research, the capabilities of LLMs to process and synthesize vast amounts of text can revolutionize the way scholars engage with literature. LLMs can consume and summarize extensive bodies of academic work, making it easier for researchers to stay abreast of developments in their field. This can be particularly beneficial for interdisciplinary research, where understanding across multiple domains is crucial.

By digesting the wealth of available academic literature, LLMs enable researchers to quickly grasp the breadth and depth of their fields, potentially uncovering connections and insights that may have otherwise gone unnoticed. This capacity to stand on the shoulders of a "broader giant"—the collective intelligence encapsulated in the corpus of academic literature—can accelerate the pace of discovery and innovation.

Furthermore, LLMs can assist in the writing process by helping researchers draft papers, generate hypotheses, or even analyze data. They can suggest alternative ways to phrase or structure arguments, identify gaps in logic or research, and ensure that the writing adheres to the conventions of academic discourse.

However, the integration of LLMs into education also necessitates careful consideration of pedagogical principles and ethical standards. The personalized learning experiences offered by AI must be aligned with educational goals and outcomes, and there must be an oversight to ensure that the use of AI supports equitable access to learning opportunities.

The potential for LLMs in education is vast, offering the promise of more inclusive, effective, and engaging learning experiences. As the technology continues to mature, it will be essential to navigate its integration into educational systems thoughtfully, ensuring that it complements and enhances human teaching and the fundamental goals of education. With the right approach, LLMs could indeed revolutionize the educational landscape, empowering learners and educators alike.

The healthcare industry, with its complex and ever-evolving body of knowledge, presents an ideal landscape for the deployment of LLMs. The benefits that LLMs can bring to the table are substantial and multifaceted, addressing some of the most pressing challenges in modern medicine.

LLMs have the capability to process and interpret vast amounts of medical literature, from research papers and clinical trial reports to patient health records. This ability to parse complex texts with high precision allows LLMs to assist medical professionals in staying current with the latest research and medical advancements without the overwhelming demand on their time typically required for such extensive literature review.

One of the most significant applications of LLMs in healthcare is their potential to assist in the diagnostic process. By analyzing patient records and medical histories, LLMs can help identify patterns and correlations that might not be immediately apparent to human practitioners. They can suggest possible diagnoses based on symptoms, lab results, and medical imaging, providing a valuable second opinion to doctors and potentially reducing diagnostic errors.

In terms of treatment, LLMs can contribute to the development of personalized medicine. By taking into account a patient's unique genetic makeup, lifestyle, and disease history, LLMs can aid in creating highly tailored treatment plans that are more likely to be effective for the individual patient. This personalization can extend to the recommendation of medications, consideration of potential side effects, and even suggestions for lifestyle adjustments that could improve the patient's health outcomes.

Moreover, LLMs can be instrumental in patient care by enhancing the communication between patients and healthcare providers. They can be used to generate patient-friendly explanations of medical conditions and treatments, thus empowering patients with a better understanding of their health and care plans.

The integration of LLMs into healthcare also has the potential to improve the efficiency of healthcare delivery. For instance, the automation of administrative tasks such as coding, billing, and the documentation of patient encounters can free up healthcare professionals to spend more time on direct patient care, thus improving the overall patient experience and outcomes.

However, the application of LLMs in healthcare does not come without challenges. The accuracy of the information provided by LLMs is of paramount importance, as mistakes can have serious, if not fatal, consequences. There are also concerns regarding patient privacy and data security, given the sensitive nature of medical records. Ensuring that LLMs are used in a way that complies with medical regulations and ethical standards is essential.

The potential for LLMs to revolutionize the healthcare industry is clear. By assisting medical professionals in diagnostics, treatment planning, and staying informed of medical research, LLMs could significantly improve patient care and outcomes. The adoption of LLMs in healthcare promises not only operational efficiencies but also advancements in personalized medicine and patient engagement. As with any transformative technology, it is crucial to approach the integration of LLMs with careful consideration of the ethical, legal, and practical implications to fully realize their benefits in the healthcare sector.

As we advance into the era of LLMs, their capabilities have prompted as much excitement as they have raised important ethical and societal questions. The core issues revolve around data privacy and the mitigation of biases that may be present in the data these models are trained on.

Data privacy emerges as a critical concern because, to achieve their advanced level of understanding and generation of human language, LLMs require training on vast datasets. These datasets typically contain information gleaned from a wide array of sources, including texts that may contain personal information. To ensure that the use of such data does not infringe on individual privacy rights, it is imperative that data used to train LLMs is sourced responsibly. This involves obtaining explicit consent from individuals whose data is being used, anonymizing data to protect individual identities, and adhering strictly to data protection regulations such as the **General Data Protection Regulation (GDPR)** in the European Union or other local data protection laws.

Beyond privacy, the issue of bias in LLMs is of paramount importance. These models learn to make predictions based on the data they are fed; if this data contains biases—whether related to gender, race, ethnicity, or socioeconomic status—the model is likely to replicate and potentially amplify these biases in its outputs. This can have serious implications, leading to discriminatory practices and reinforcing societal inequalities. For instance, if a model trained on historical hiring data that exhibits gender bias is used to screen job applicants, it may inadvertently continue to favor one gender over another.

Acknowledging these biases is the first step, but actively working to address them is crucial. This can involve curating training datasets to be as diverse and representative as possible, implementing algorithmic checks that identify and mitigate bias, and continuously monitoring the output of LLMs to ensure that they do not propagate prejudiced viewpoints.

Another aspect to consider is the interpretability of LLM decisions. As these models become more complex, understanding the rationale behind their predictions becomes more challenging. There is a growing need for transparency in AI decision-making, especially as LLMs begin to play a role in critical areas such as healthcare, law, and finance. Developers and stakeholders must strive for models that are not only effective but also interpretable so that their decisions can be understood and trusted by users.

Furthermore, as LLMs become more integrated into our daily lives, the question of job displacement arises. While LLMs can augment human capabilities in many ways, there is also the potential for them to replace jobs in fields such as customer service, content creation, and more. This calls for a proactive approach to reskilling and education, ensuring that the workforce is equipped to work alongside AI and leverage its capabilities rather than be sidelined by it.

Transparency in the decision-making of LLMs is a critical issue that grows in importance as these models become more integrated into decision-making processes that significantly affect individual lives. The ability to understand and scrutinize the reasoning behind an LLM's output is not just a matter of establishing trust but also one of ensuring accountability.

Trust is a cornerstone of any technology's adoption, and when it comes to AI, the users' trust hinges on their understanding of how the AI arrives at its conclusions. If the workings of an LLM are opaque, it undermines the confidence that users—and society at large—might have in it. Trust can be bolstered when there is clarity and openness about the internal mechanisms and logic that guide the LLM's responses and decisions.

Fairness in AI decision-making is closely tied to transparency. If an LLM were to make a decision that has an unfair or discriminatory impact, it's essential to be able to dissect the decision pathway to identify the origin of the bias. Transparency enables the detection and correction of any such biases, ensuring that the models operate in a manner that is equitable and just.

Accountability is another facet of transparency. When an LLM's decision has a significant impact on a person's life, such as in legal sentencing, loan approval, or job application screening, the ability to trace and understand the decision process is vital. If an outcome is negative or even harmful, stakeholders must be able to hold the appropriate parties responsible, and this is only possible if the decision-making process is transparent.

Compliance with regulations is also a driving force behind the need for transparent AI. Regulatory frameworks such as the European Union's GDPR mandate that individuals have the right to understand the logic behind automated decisions that affect them. This legal requirement makes it imperative for LLMs to operate in a transparent manner, providing clear explanations for their outputs.

Lastly, the improvement and refinement of LLMs rely on the ability to interpret their decision-making processes. If the rationale behind a model's decision is not clear, it becomes much more difficult to identify mistakes, learn from them, and, ultimately, improve the model's performance. Developers need this insight to refine and advance the model's capabilities.

Despite the recognized need for transparency, the complexity of LLMs often leads to a "black box" scenario where even the creators of the model may not fully understand why a certain decision was made. This is especially true for models based on deep learning, which can involve millions of parameters and complex data representations.

The field of **explainable AI** (**XAI**) aims to address this issue by developing models that inherently provide more interpretable decision-making processes or by creating tools to decipher the decisions of existing models. XAI is an area of active research, seeking to bridge the gap between AI performance and human comprehension.

In essence, the push for transparency in LLMs is a multifaceted effort that encompasses the ethical, practical, and regulatory dimensions of AI deployment. As the presence of LLMs in critical sectors grows, so too does the imperative for these systems to be as open as they are intelligent, ensuring that their integration into our lives and livelihoods is characterized by trust, fairness, and accountability.

The development and deployment of LLMs carry with them a responsibility to bridge rather than widen the existing digital divide. It is essential that these advanced technologies are made inclusive and accessible across all sectors of society to ensure that the benefits they provide are not limited to a privileged few but are instead shared equitably.

Inclusivity in the context of LLMs means that the models must be trained on diverse datasets that reflect the breadth of human experience and language. They should not solely represent the most dominant languages or dialects but also encompass a variety of sociolects, ethnolects, and regional vernaculars. This inclusivity in training data helps ensure that the model's utility is not confined to a subset of the population but is valuable and usable by people from different linguistic, cultural, and social backgrounds.

Accessibility is equally crucial. LLMs should be designed with user interfaces that are easy to navigate for people with varying levels of digital literacy. Additionally, considerations for individuals with disabilities should be integrated into the design process to ensure that these tools are usable by everyone, including those who require assistive technologies to interact with digital platforms.

The deployment of LLMs also needs to take into account the varying levels of technology infrastructure across different regions. Efforts should be made to ensure that LLMs do not require prohibitively high computational resources or connectivity bandwidth, which could otherwise limit their use to areas with more advanced technological infrastructure. Cloud-based solutions and adaptive technology can help in making LLMs more widely available, irrespective of local hardware limitations.

Moreover, the potential benefits of LLMs—such as enhanced learning opportunities, streamlined access to information, and improved efficiencies in workplaces—should be disseminated widely. This means not only making the technology itself accessible but also providing the necessary education and support to enable individuals and communities to leverage these tools effectively.

The threat of the digital divide growing as a result of advancements in AI is real. As LLMs become more pervasive in areas such as education, employment, and access to services, there is a risk that those without access to these technologies will fall further behind. To combat this, governments, educational institutions, and industry leaders must collaborate to create initiatives that promote digital inclusion. This could involve investing in infrastructure, providing training programs to upskill the workforce, and ensuring that educational curricula evolve to impart the necessary skills to engage with AI technologies.

Interdisciplinary collaboration emerges as a fundamental strategy in addressing the multifaceted challenges and opportunities presented by LLMs. The future of these advanced AI systems is not solely in the hands of technologists who build and refine them; it also rests on the insights and oversight of professionals from a variety of other fields.

Linguists play a crucial role in this collaborative effort. Their expertise is invaluable in training LLMs to understand and generate natural language in a way that is accurate, culturally sensitive, and contextually appropriate. They can provide guidance on the nuances of language that are often lost in translation between different languages and dialects, ensuring that LLMs serve a broader user base.

Ethicists are essential in steering the development of LLMs toward a path that aligns with moral principles and societal values. Their involvement ensures that considerations such as fairness, privacy, and the potential for bias are factored into the development process from the ground up. They can help foresee ethical dilemmas and work toward preemptive solutions.

Legal experts contribute by ensuring that the development and application of LLMs comply with existing laws and regulations. They can also anticipate the need for new legal frameworks to address the novel questions that LLMs introduce, such as intellectual property rights in the context of machine-generated content or liability issues stemming from AI-driven decisions.

Policymakers have the responsibility to create an environment that fosters the responsible growth of LLM technologies. This includes formulating policies that encourage innovation while protecting the public from potential harm. Policies might include funding for research into the impacts of LLMs, regulations that promote transparency and accountability, or initiatives that address the digital divide.

The collaborative approach extends beyond creating frameworks for the ethical development and deployment of LLMs. It is also about ensuring that the technology is applied in ways that maximize its potential for social good. For example, technologists can develop the LLMs, linguists can ensure they communicate effectively, ethicists can oversee the moral implications, legal experts can navigate the regulatory landscape, and policymakers can implement these technologies to serve the public interest.

The societal benefits of LLMs are significant—they can revolutionize industries, enhance productivity, and open up new avenues for innovation. However, these benefits come with risks, such as the potential for exacerbating social inequalities or infringing upon individual rights. An interdisciplinary approach provides a holistic view of these technologies, taking into account the varied implications of their deployment.

The revolution brought about by LLMs is setting the stage for a profound transformation in the way we engage with the digital world. This paradigm shift is not limited to technological advancements; it heralds a new chapter in human-computer interaction, learning, and the very fabric of the workforce. As we navigate the waters of this new era, it is crucial that we chart a course that leverages these advancements for the benefit of society as a whole.

The impact of LLMs on the workplace is already becoming evident. Tasks that once required extensive human effort are now being augmented or even replaced by intelligent systems capable of analyzing data, generating reports, and managing information with a speed and accuracy that far exceed human capabilities. In the future, this transformation will enable human workers to concentrate on tasks that demand creativity, critical thinking, and emotional intelligence—areas where human skills excel and AI has yet to make inroads.

In the realm of education, LLMs offer the potential for personalized learning experiences that adapt to the individual learner's pace, style, and interests. This could democratize education, making high-quality, tailored learning accessible to students around the globe, regardless of geographical or socioeconomic barriers. The implications for lifelong learning and the continuous upskilling of the workforce in response to an ever-changing job market are profound.

Our interactions with the digital world are also being redefined. LLMs facilitate more natural and intuitive interfaces, allowing us to communicate with digital systems as we would with another human being. This enhances accessibility, breaking down barriers for those who may have previously found technology intimidating or inaccessible.

However, the trajectory of the LLM revolution must be guided by a clear and deliberate vision that takes into account the ethical, social, and economic implications of these technologies. Ethically, we must ensure that LLMs are developed and deployed in a manner that respects privacy, minimizes bias, and promotes fairness. Socially, we must be vigilant to ensure that the benefits of LLMs do not exacerbate existing inequalities but instead provide opportunities for all. Economically, we need to ensure that the efficiency gains brought by LLMs do not come at the cost of job displacement but are instead channeled toward economic growth that benefits society at large.

As we step into this new era, it is imperative that we do so with a commitment to ensuring that the LLM revolution fosters a future that is informed by the highest standards of ethical practice, that operates with an efficiency that enhances rather than undermines human endeavor, and that commits to equity, ensuring that the fruits of this revolution are enjoyed by all members of society. If we navigate this revolution with foresight and responsibility, we have the opportunity to shape a future that is not only technologically advanced but also socially and economically inclusive.

Index

A

Other Books You May Enjoy

If you enjoyed this book, you may be interested in these other books by Packt:

Mastering NLP from Foundations to LLMs

Lior Gazit, Meysam Ghaffari PhD

ISBN: 978-1-80461-918-6

- Master the mathematical foundations of machine learning and NLP
- Implement advanced techniques for preprocessing text data and analysis Design ML-NLP systems in Python
- Model and classify text using traditional machine learning and deep learning methods
- Understand the theory and design of LLMs and their implementation for various applications in AI
- Explore NLP insights, trends, and expert opinions on its future direction and potential

Building Data-Driven Applications with LlamaIndex

Andrei Gheorghiu

ISBN: 978-1-83508-950-7

- Understand the LlamaIndex ecosystem and common use cases
- Master techniques to ingest and parse data from various sources into LlamaIndex
- Discover how to create optimized indexes tailored to your use cases
- Understand how to query LlamaIndex effectively and interpret responses
- Build an end-to-end interactive web application with LlamaIndex, Python, and Streamlit
- Customize a LlamaIndex configuration based on your project needs
- Predict costs and deal with potential privacy issues
- Deploy LlamaIndex applications that others can use

Packt is searching for authors like you

If you're interested in becoming an author for Packt, please visit `authors.packtpub.com` and apply today. We have worked with thousands of developers and tech professionals, just like you, to help them share their insight with the global tech community. You can make a general application, apply for a specific hot topic that we are recruiting an author for, or submit your own idea.

Share Your Thoughts

Now you've finished *Decoding Large Language Models*, we'd love to hear your thoughts! Scan the QR code below to go straight to the Amazon review page for this book and share your feedback or leave a review on the site that you purchased it from.

`https://packt.link/r/1-835-08465-6`

Your review is important to us and the tech community and will help us make sure we're delivering excellent quality content.

Download a free PDF copy of this book

Thanks for purchasing this book!

Do you like to read on the go but are unable to carry your print books everywhere?

Is your eBook purchase not compatible with the device of your choice?

Don't worry, now with every Packt book you get a DRM-free PDF version of that book at no cost.

Read anywhere, any place, on any device. Search, copy, and paste code from your favorite technical books directly into your application.

The perks don't stop there, you can get exclusive access to discounts, newsletters, and great free content in your inbox daily

Follow these simple steps to get the benefits:

1. Scan the QR code or visit the link below

https://packt.link/free-ebook/978-1-83508-465-6

2. Submit your proof of purchase
3. That's it! We'll send your free PDF and other benefits to your email directly